Revolutionary Therapies

How the California Stem Cell Program Saved Lives, Eased Suffering — and Changed the Face of Medicine Forever

Other World Scientific Titles by the Author

California Cures!
How the California Stem Cell Program is Fighting Your
Incurable Disease!
ISBN: 978-981-3231-36-8 (hardcover)
ISBN: 978-981-3270-38-1 (paperback)
ISBN: 978-981-3231-37-5 (ebook for institutions)
ISBN: 978-981-3231-38-2 (ebook for individuals)

Stem Cell Battles: Proposition 71 and Beyond
How Ordinary People Can Fight Back against the Crushing Burden of
Chronic Disease — with a Posthumous Foreword by Christopher Reeve
ISBN: 978-981-4644-01-3 (hardcover)
ISBN: 978-981-4618-27-4 (paperback)
ISBN: 978-981-4618-28-1 (ebook for institutions)
ISBN: 978-981-4618-29-8 (ebook for individuals)

Revolutionary Therapies

How the California Stem Cell Program Saved Lives, Eased Suffering — and Changed the Face of Medicine Forever

Don C. Reed
Americans for Cures Foundation, USA

World Scientific

NEW JERSEY · LONDON · SINGAPORE · BEIJING · SHANGHAI · HONG KONG · TAIPEI · CHENNAI · TOKYO

Published by

World Scientific Publishing Co. Pte. Ltd.

5 Toh Tuck Link, Singapore 596224

USA office: 27 Warren Street, Suite 401-402, Hackensack, NJ 07601

UK office: 57 Shelton Street, Covent Garden, London WC2H 9HE

Library of Congress Control Number: 2019058297

British Library Cataloguing-in-Publication Data
A catalogue record for this book is available from the British Library.

REVOLUTIONARY THERAPIES
How the California Stem Cell Program Saved Lives, Eased Suffering — and Changed
the Face of Medicine Forever

ISBN 978-981-121-328-1 (hardcover)
ISBN 978-981-121-329-8 (ebook for institutions)
ISBN 978-981-121-330-4 (ebook for individuals)

For any available supplementary material, please visit
https://www.worldscientific.com/worldscibooks/10.1142/11638#t=suppl

Dedication

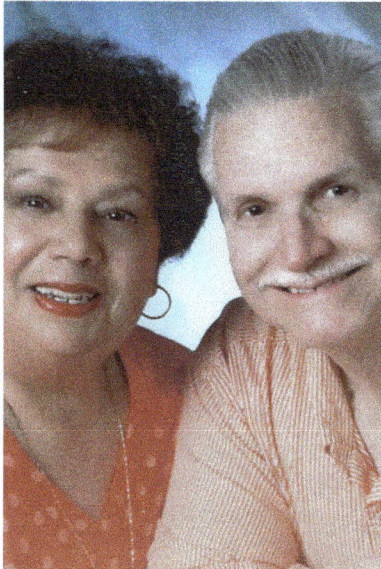

To Gloria:

From the moment I saw you in your white lace wedding dress, framed in a beam of sunlight, at the church door, I have never ceased to marvel at my good luck in being married to you.

You helped me get my scuba-diving job at Marine World, doing the interview for me, although you were 7½ months pregnant! The head diver said, "Lady, don't have your baby here, I'll give him the job!"

You tricked me into becoming a teacher, making me take dozens of courses, one at a time. I never understood why, until you said, "Now is when you become a teacher — or I divorce you!" (You said you were only joking, but I am not so sure — and I did become a teacher!)

You even told me when it was time to "retire" and work full-time on stem cell research advocacy.

At every turning point in my life, you were there, to counsel and guide me, to brighten our lives.

As the mother of my children, my best friend, my love, you are the reason for my hope in heaven — you are the sunshine of my life.

Thank you for being you.

With joy
Don C. Reed

Contents

Introduction: The Odds Against the California Stem Cell Program

Governor Arnold Schwarzenegger authorizes a loan of $150 million to officially open the California Institute for Regenerative Medicine. At left is CIRM author and Chair Emeritus Bob Klein; on right is author/advocate Don C. Reed. (Photo by Viartis.com.)

From the beginning, the odds were steep against CIRM, the California stem cell program, technically called the California Institute for Regenerative Medicine (CIRM). A stem cell research program asking for three *billion* dollars?

But leading the charge was Bob Klein, an expert in raising funds for good causes.

He had previously developed the California Housing Finance Authority, which offered low-cost housing loans to needy families — not one dime of which went to his own company, Klein Financial Enterprises, Inc. He also helped raise one and a half billion federal dollars for diabetes research. But his greatest adventure was Proposition 71, the California Stem Cells for Research and Cures Act.[1]

Prop 71 was a massive group effort, backed by dozens of medical and education groups, hundreds of patient advocate families who gathered 1.1 million petition signatures — and 5.9 million voters who said "YES!" to stem cell research.

We won on the ballot, but still the fight continued.

Multiple lawsuits tried to shut us down, and it took nearly three years to defeat them all. But in 2007, it became clear our program would prevail no matter what the ultra-conservatives threw at us. Governor Arnold Schwarzenegger cheerfully approved a loan of $150 million to let CIRM get started — and for the next 15 years, CIRM has fought to find cures for chronic disease and disability.

Which brings us to the crucial question: **should the California stem cell program be renewed?**

Or, when its funding runs out, (as it soon will), is that the end of CIRM?

This book is my answer to that question.

---

[1] https://en.wikipedia.org/wiki/Robert_N._Klein_II

1 The Silent Hurricane

Hurricane Harvey killed 70 people; American diabetes takes the lives of roughly 76,400 — every year. (Photo by CNN.com.)

As this is written, America is glued to our TV sets, watching in fascinated horror as a procession of hurricanes moves in from the sea, wreaking havoc and destruction: rain pounds and waves surge onto Texas, Puerto Rico and Florida — and more devastation is building offshore, circling in, on the way.[1]

But what if there was a silent hurricane?

In a way, there already is: making no noise, gathering few headlines, but wreaking destruction onto people's lives — in every state of the Union and across the world.

Its winds will not subside, its torrential blasts will not end; it is forever, unless we stop it.

That silent storm is chronic disease. How many people are affected by it?

---

[1] http://tinyurl.com/y28j3czb

According to the National Health Council:

"Generally incurable and ongoing, chronic diseases affect approximately 133 million Americans...more than 40% of the total population of this country."[2]

That is nearly one in two people: family members, loved ones, maybe you, definitely me. (I have peripheral neuropathy, nerve pain in the feet. Also I am a cancer "survivor", which means it may or not return.)

Consider just one chronic disease: diabetes, which last year cost America nearly 249 billion dollars, a quarter of a trillion bucks; almost double the $125 billion expense of Hurricane Harvey.[3]

Harvey has killed about 70 people: agony for the families involved.

But American diabetes has an annual death rate a thousand times higher — 76,400 per year. Internationally? 3.4 million people die each year from diabetes.[4]

Injuries? In America alone, more than 73,000 people suffer amputations of toes and feet, as a direct consequence of diabetes. Diabetes is also the leading cause of adult onset blindness, coma, and kidney failure — and that is all from just <u>one</u> chronic disease![5]

How do we fight hurricanes, those great swirling storms? Mostly, unfortunately, we don't. We try to get the people out of the path of destruction, strengthen our seawalls; but too little is done to battle the causes of hurricanes, the increase in global warming: climate change. To solve the problem, we must tackle pollution.

And diabetes? Patients get insulin injections, blood tests, and the constant monitoring of food intake — but nothing to bring about cure.

Fighting for cure is what the California Institute for Regenerative Medicine (CIRM) is about. It is not enough to maintain people in misery, we want them well.

Here is one of several possibilities being investigated by CIRM:

Imagine a plastic device, about the size of your index finger. It contains microscopic materials, *progenitors*, which will become stem cells. We hope it will function like a normal pancreas, providing insulin and other life-saving benefits.

---

[2] http://tinyurl.com/yykgfev8
[3] https://coast.noaa.gov/states/fast-facts/hurricane-costs.html
[4] https://www.cdc.gov/nchs/fastats/diabetes.htm
[5] https://nei.nih.gov/health/diabetic/retinopathy

Lives were changed by CIRM: the little girl whose life was saved; a young man fighting paralysis; a man with lung cancer, another defeating an immune disorder — and smiling behind them, Senator Art Torres, former President Randy Mills, and current Chairman of the Board, Jonathan Thomas. (Photo by CIRM blog.)

The idea is to put the device under the skin, and then — forget about it. This idea was invented by a company called ViaCyte, with research and clinical tests assisted by CIRM to the tune of roughly $51 million dollars.[6]

It is in clinical trials: being tested on volunteers as you read this.[7]

And that is why CIRM exists: to fight the silent hurricane of chronic disease.

---

[6] https://www.cirm.ca.gov/our-progress/institutions/viacyte-inc
[7] http://tinyurl.com/y3wrgg47

2 Uncle Ben's Kidneys

Andy McMahon wants to develop living "nephrons" tissue tubes to filter urine and prevent death by kidney failure. (Photo by Pressroom, USC.)

"You should get a computer!" said my cousin Tom Snyder, a couple of decades ago, "It would help you write your books!"

"Sure," I snapped, "Next time I get an extra $5,000 (computers cost a lot back then) I will just dash right out and get one!"

Months passed, and I forgot about Tom's seemingly casual remark.

Then suddenly, knocking at the door, was my cousin. He had a big beaming smile on his face and a box in his arms: a Macintosh computer.

Tom had organized my family to raise the money. My Uncle Ben was one of the primary funders of this greatest material gift of my life.

I wish I had a happy ending for this story.

But Ben Snyder had End Stage Renal Failure: his kidneys ceased to function. No longer would they filter his blood, transferring the poisonous waste into the urine, to then pass safely from the body; instead the toxins accumulated.

He had to endure hemodialysis: blood washing. Three times a week he went to the hospital. Needles stabbed into his arm, and a massive amount of blood (60–70% of all he owned) was passed through a filter, cleaned and put back.

The procedure took three hours.

It kept him alive for several years, but then he died: a good man, taken too soon.

Today, kidney disease affects a staggering number of people: thirty million in America alone. Many more may have it, but are not yet diagnosed.[1]

And the end-stage condition which took my Uncle's life?

"Over 650,000 Americans suffer ... (the) life-threatening condition caused by the loss of kidney function."

— Kevin McCormack, CIRM Press Release, January 18, 2018.

So, what's to be done?

CIRM, the California Institute for Regenerative Medicine, is tackling kidney disease from several directions.

Andy McMahon, a developmental geneticist, is studying the natural development of the kidney. By studying its biology, he hopes to learn how a healthy kidney repairs itself.[2]

Have you heard of "organoids"? These man-made mini-organs may provide a great avenue for drug testing — instead of trying a new therapy on a person, we can test it on a glob of tissue, with no risk to

---

[1] https://www.kidney.org/news/one-seven-american-adults-estimated-to-have-chronic-kidney-disease

[2] https://stemcell.keck.usc.edu/mcmahon-named-university-professor-davies-mataric-and-pastor-named-distinguished-professors/

people. This could speed up the safety and efficacy trials, and lower the costs of new drugs.

Down the road, some feel, organoids may be used as replacements for various parts of the body. But whether for testing (disease in a dish) or replacement, the organoid must be "relevant" (similar to the organ being studied), as Andy said in a recent phone interview.

A key element of the kidney is the nephron, a microscopic human tube, which helps filter waste. Dr. McMahon wants to develop nephrons (perhaps a million per kidney) and learn how to protect or replace them.[3]

Another approach is transplantation. When a kidney donor can be found, (difficult but not impossible) the new kidney has to be installed and accepted by its new owner's body. To prevent rejection, immune-suppressants must be used. These can bring danger — infections, heart disease, cancer, diabetes, all manner of bad stuff.

However, a CIRM-funded company, Medeor Therapeutics, may have the answer. Injecting stem cells and immune cells from the donor may make his/her transplant organ "fit in," and not be rejected.[4]

"Medeor's stem cell-based therapy aims to prevent transplant rejection, and eliminate the need for immunosuppression..." said CIRM President Maria T. Millan... "If they are successful, this approach could be developed for other organ (transplants) including heart, liver, and lung..."

Still another approach, this one by Humacyte, Inc., will use stem cells to make the dialysis process safer, using a bioengineered vein. Instead of a plastic tube inserted into the arm, with accompanying risk of infection, a "bioengineered vein (made of stem cells) is implanted in the arm and used to carry the patient's blood to and from their body during dialysis."[5]

The tube would gradually morph into the patient's body, as if it had been born there.

But whether repair or replacement, new nephrons or biomedical tubing, something major must be done.

Because with one American in seven suffering kidney disease, almost everybody has an Uncle Ben.

---

[3] https://www.cirm.ca.gov/our-progress/awards/repair-and-regeneration-nephron
[4] https://blog.cirm.ca.gov/2018/01/22/cirm-invests-in-medeor-therapeutics-phase-3-clinical-trial-to-help-kidney-transplant-patients/
[5] https://www.ncbiotech.org/news/humacyte-gets-10m-boost-clinical-trial

3 Blindness for the Old

"Our CIRM disease team...aims to treat (adult blindness) through replacement of cells..." (Photo by The Stem Cellar.)

Do you worry about going blind? If so, you are not alone. For eight million Americans with Age-related Macular Degeneration (AMD), such concerns are valid.

When you first notice it, AMD is just a dot in the middle of your vision: an annoyance, something you can't quite blink away. But the dot grows until it is like someone holding a thumb over your eye. You can see just a little bit, around the edges. Worse is to come. The ability to see colors leaves you, followed by the loss of night vision. Finally you cannot read, nor see faces, nor drive a car safely.

That is "dry" macular degeneration, and it's bad. But it may also lead to the "wet" version (cells leaking blood inside the eye), which can bring total blindness.

As vision weakens, it becomes easy for the elderly to fall and break bones. Depression sinks in, sometimes accompanied by suicidal thoughts.

Medical help exists (injections in the eyes), but it is expensive, and can only slow the advance of the disease; it does nothing to bring healing.

The California stem cell program wants to protect and restore your vision.

Think of the back of your eyeball. Inside it there is the macula (rhymes with Dracula) which provides central vision, color, and night sight.

The macula's nourishment is provided by a thin layer of specialized cells, the Retina Pigment Epithelial (RPE) cells.

Without the RPE, the macula starves, and the eye goes blind.

So — the California stem cell program is attempting to build new RPE cells.

A team of scientists, led by Drs. Mark Humayun (University of Southern California) David Hinton (also from USC), Dennis Clegg (UC Santa Barbara), and Pete Coffey (Almac Diagnostics, Inc., UK) is working with embryonic stem cells to restore vision.[1]

Using microscopic dots of tissue left over from the In Vitro Fertilization procedure (which would otherwise be thrown away), the scientists are developing RPE cells, to provide nutrients to the damaged eye.

Tested first on rodents and pigs, the new therapy was not only safe, but provided some measure of positive results.

Next, the procedure will be tried on people: first for safety, and then (in small but increasing doses) to see if it might slow the vision loss, or even reverse it.[2]

How important could that be, financially?

"Dry mac" alone is estimated to cost America about $24 billion a year — eight times the cost of our entire stem cell program.

But when we consider *the world's expense of vision loss* — from all sources?

Estimates run as high as three *trillion* dollars — ($2,954 billion). Stacked up in dollar bills, that thirty tons of money could almost reach the moon.[3]

---

[1] https://www.cirm.ca.gov/our-progress/awards/stem-cell-based-treatment-strategy-age-related-macular-degeneration-amd

[2] https://www.cirm.ca.gov/our-progress/awards/phase-1-safety-assessment-cpcb-rpe1-hesc-derived-rpe-cell-coated-parylene

[3] http://retinatoday.com/2010/04/estimates-of-global-cost-of-vision-loss-at-3-trillion

Listen to Dr. Humayun speaking about his step-by-step procedures:

"Our CIRM disease team, which we call 'The California Project to Cure Blindness', aims to treat AMD through replacement of (non-working) cells ... we have been able to produce cells that exhibit many characteristics of normal ... cells ...

"... animal studies demonstrated the safety of the technique as well as the restoration of some functional vision...Hopefully, this will lead us to successful ... trials with sight restoration in cases currently untreatable."[4]

And when at last it was tried with people? In a small human clinical trial:

"Of five patients enrolled in the Phase 1/2a trial, four maintained their vision in the treated eye, two showed improvement in the stability of their vision, and one patient had (substantial) improvement in vision on a reading chart. There were no serious side effects or unanticipated problems."[5]

There are no guarantees in science, of course. But with reliable funding, they have a chance. This is why the *long-term funding* of the California stem cell program is so valuable; it allows time to attack blindness like the large-scale menace it is.

Humayun now seeks a larger trial, probably around 15 people. The first four clients will be followed, to track their progress.

And that is only one team. Visit the CIRM website on blindness and see some of the other approaches: see how $141 million ($140,914,431) is dedicated to the fight against vision loss.[6]

---

[4] https://www.cirm.ca.gov/our-progress/awards/phase-1-safety-assessment-cpcb-rpe1-hesc-derived-rpe-cell-coated-parylene

[5] https://www.cirm.ca.gov/about-cirm/newsroom/press-releases/04042018/stem-cell-agency-funded-clinical-trial-age-related

[6] https://www.cirm.ca.gov/our-progress/disease-information/blindness-fact-sheet

4 So, You Want to be a Stem Cell Scientist?

Want to spend your Summer in a stem cell lab? (CIRM photo.)

Do you want to become a stem cell scientist, or know someone who does?

If so, this chapter is a gift for you; or, more accurately, three gifts — all from the California Institute for Regenerative Medicine (CIRM).

<u>First</u>, if you are in High School or college, would you like your science teacher to teach a stem cell program? If so, there is a free stem cell curriculum ready and waiting for her/him to teach. This was developed with the cooperation of CIRM as part of Senate Bill 471 (Romero, Steinberg and Torlakson), the California Stem Cell and Biotechnology Education and Workforce Development Act of 2009.

It offers a wide variety of teaching materials, a full curriculum:

"On this site, you will be able to: explore opportunities for teaching your students about stem cell science; find resources for student projects;

locate a scientist to discuss stem cell science with your class; get involved in the growing areas of stem cell research and biotechnology."[1]

Second, for high school students, here is an amazing possibility.

How would you like to spend a Summer working in a stem cell research lab?

SPARK: The Summer Program for Advancing Regenerative Medical Knowledge offers you that chance.

SPARK "offers California high school students (the) opportunity to gain hands-on training in stem cell research at some of the leading research institutes in California. The...program specifically selects students who represent the diversity of California, particularly those who might not otherwise have opportunities...due to socioeconomic constraints."[2]

The third gift is the BRIDGES program, which may help you answer the age-old employer question: "What experience do you have?"

Nobody wants to whine "How can I get experience, when nobody will hire me?"

Instead, how about: "I worked all year at a world-renowned biomedicine lab" — for which you got paid $30,000!

Here's how it works: college students take basic stem cell training at their home college for the first year, followed by a second year in a lab position at one of the research institutions taking part in the program. For this they will receive a stipend, up to $30,000, as well as hands-on experience working at the lab.

More than 1,100 students have completed the Bridges program since its 2009 inception. Half are working full time in biomedicine; others are in further education.

The Bridges program shapes the future.

At San Diego State University, for example: "Highly competitive trainees will be recruited from the university's diverse student populations which include individuals that might not otherwise have the chance to attain the essential expertise to contribute to the ultimate goal of delivering stem-cell based therapies to patients."[3]

Want to know more? Email info@cirm.ca.gov.

---

[1] http://www.cirm.ca.gov/our-impact/education/stem-cell-portal
[2] https://blog.cirm.ca.gov/tag/spark/
[3] https://www.cirm.ca.gov/our-progress/awards/bridges-stem-cell-research-internship-program

5 The One-Leg Placebo

"Ask an Expert" about the battle against ALS, sometimes called Lou Gehrig's disease, long considered incurable — but maybe not forever. (CIRM photo.)

Visualize a high school, with 1,000 students: all deathly sick; their limbs failing, muscles growing weak and paralyzed; they will die in 3–5 years. Every student has Amyotrophic Lateral Sclerosis (ALS), sometimes called Lou Gehrig's disease.[1]

Now make that six such schools, 6,000 sufferers — that is how many Americans have the slow death sentence of ALS. In California alone, a person dies of this disease every one and a half days.

CIRM is fighting on their side.

---

[1] https://www.cirm.ca.gov/our-progress/disease-information/amyotrophic-lateral-sclerosis-als-fact-sheet

To better understand the enemy, view a cartoon explanation of ALS from Americans For Cures Foundation (for which I work) at the below URL.[2]

Today, a company called BrainStorm Therapeutics Inc. is beginning a Phase 3 set of clinical trials to attack ALS.

Led by veteran neuroscientist Dr. Ralph Kern (and supported by a $15.9 million grant from the California stem cell program), they have a stem cell product called NurOwn®.

"One hundred people will receive NurOwn®. This technology takes … stem cells, and … converts them into biological factories secreting a variety of neurotrophic factors … for the growth and survival of developing neurons." — Mary Kay Turner, Vice President, Patient Advocacy, Brainstorm Cell Therapeutics Ltd.

Sixty of those patients will be in California, at UC Irvine and the California Pacific Medical Center. (CIRM can only pay for research and therapies administered in our state.) California's City of Hope Center for Biomedicine and Genetics will produce the "clinical supplies of NurOwn® for the Phase 3 clinical study…"[3]

Here is Dr. Kern, on his company's approach to ALS:

"We are at a point where Central Nerve System science and technology is coming of age — our technology, NurOwn®, has been developed and advanced for over 10 years. We're now conducting a pivotal multicenter phase 3 trial at 6 sites in the US, which makes it the most advanced stem cell therapy approach currently in development for ALS…We are hoping to quickly advance the phase 3 program and bring a much needed solution to ALS patients and their families."

Another clinical trial is being led by Dr. Clive Svendsen at Cedars-Sinai Medical Center:

"This project will transplant stem cells secreting the powerful growth factor GDNF into the spinal cords of patients with (ALS) (to) delay motor neuron death and thus treat the disease."[4]

They have an amazing way of testing their therapy: a one-leg placebo.

---

[2] https://americansforcures.org/disease/als/

[3] http://ir.brainstorm-cell.com/phoenix.zhtml?c=142287&p=RssLanding&cat=news&id=2293816

[4] https://www.cirm.ca.gov/our-progress/awards/progenitor-cells-secreting-gdnf-treatment-als

Remember your high school science: the placebo effect, where a person thinks they are getting better, and may even improve for a while, because of that belief?

To prevent that, a "blind" is established in the trials. This is someone who seems to be part of the trial, and goes through everything, except the medication.

That always seemed unfair to me, because what if the treatment was good? The person receiving the placebo might be cheated from a chance to get well.

But in Svendsen's trial, each patient is their own placebo effect — because the cells are administered to the part of the spine which controls the movement of one leg.

Every person in the trial has one medicated leg, and one not — and they do not know which leg is which. So if one leg gets better, and that is the leg that got the stem cells — then you have something. If it was the leg that did not get the treatment, then it could be a placebo effect. (If the experiment works, as I understand it, those who got the placebo first will get the real stuff later.

If you have ever been to a board meeting of CIRM (and you should, you are welcome — just go to the website, and click on "Meetings" for where and when) you may have noticed a small, seemingly fragile woman, Dianne Winokur.[5]

Ms. Winokur has been fighting to end ALS for more than two decades: ever since she lost two sons to the condition, Douglas and Hugh.[6]

She does not talk much, but when she does, it counts. Here is what she said, after the decision was made to fund Dr. Svendsen:

"In the 150 years since ALS was first described, there has been little progress in finding therapies. In the last few years, thanks to new technologies, increased interest, and CIRM support, we finally seem to be seeing some encouraging signs in the research. Dr. Svendsen has been at the forefront of this effort for the 20 years I have followed his work. I commend him, Cedars-Sinai, and CIRM."

"On behalf of those who have suffered through this cruel disease, and their families and caregivers, I am filled with hope."

---

[5] https://blog.cirm.ca.gov/tag/diane-winokur/
[6] https://blog.cirm.ca.gov/2016/10/24/ingenious-cirm-funded-stem-cell-approach-to-treating-als-gets-go-ahead-to-start-clinical-trial/

6 Can We Lower the Prices of Medicine and Therapies?

Geoff Lomax specializes in the Alpha Sites and and the fight to lower costs of medicine and therapies. (Twitter Photo.)

Why do new drugs and therapies cost so much?

Part of the reason is greed. The love of profits motivates some people with a lust stronger than sex or donuts.

Drug magnates like Martin Shrekeli apparently have an unquenchable lust for more and more money, no matter who suffers. When he raised the price of a single pill from $13.50 to $750.00, (thirteen dollars to seven hundred fifty bucks?) he established a new standard for greed.[1]

But there is also a logistical problem, leading to high drug cost. In biomed it is called the "Valley of Death." It is not a physical place, where new drugs or therapies go to die: just a too-long series of tests required

[1] https://www.nytimes.com/2015/09/21/business/a-huge-overnight-increase-in-a-drugs-price-raises-protests.html?_r=0

by the Food and Drug Administration (FDA) before a new medical product or practice can be sold.

How much does the testing cost to bring one new drug or therapy through the "valley of death" of FDA testing? You may not believe it.

Costs may be as high as $3.9 **billion** dollars to take one drug through testing.[2]

If a company can't raise that mountain of money, the new drug or therapy may die.

Don't get me wrong, I respect the FDA. We need the good work they do. Without careful tests, anything could be passed off as a stem cell therapy, and no one would know what was being put inside our bodies.

An untested so-called "Medicine" might make people worse. Did you hear about the three ladies who paid to have adult stem cells made from their own fat tissues and injected into their eyes, and the "treatment" made them all blind?[3]

Even so, we need the research to go forward at full speed. That isn't happening.

One stem cell therapy (Hans Keirstead's ground-breaking paralysis work, with early research paid for by the Roman Reed Spinal Cord Research Act of 1999) has been in testing for two decades. It is good stuff; restoring upper body control to several newly paralyzed patients.[4]

But the testing process is too long and too expensive; we must find a better way.

And maybe, CIRM does.

What if, instead of having to design test after test after test — each an original and therefore expensive — we had an FDA-approved place to get the tests done, all set up, ready to go, like plug and play for a computer program?

That is the thinking behind the *Alpha Clinics*.

The California stem cell program has developed five "Alpha Sites" — at UC San Diego, UCLA/UCI, the City of Hope, UC Davis and the University of California at San Francisco. There is also a stem cell center

---

[2] https://www.scientificamerican.com/article/cost-to-develop-new-pharmaceutical-drug-now-exceeds-2-5b/

[3] https://www.washingtonpost.com/news/to-your-health/wp/2017/03/15/three-women-blinded-by-unapproved-stem-cell-treatment-at-south-florida-clinic/?utm_term=.9bf96deff07b

[4] https://blog.cirm.ca.gov/2016/09/07/young-man-with-spinal-cord-injury-regains-use-of-hands-and-arms-after-stem-cell-therapy/

to coordinate the storage of new information, and assist patients and physicians with their questions.

Each "Alpha Site" reaches a different patient community, hopefully allowing people from all across the state to have access to stem cell clinical trials closer to them. UCSF, for instance, will serve Northern California and the Bay Area.

Will the Alpha Stem Cell Network accelerate progress toward curing chronic diseases and disabilities?

They are doing it right now.[5]

Here are some of the conditions in clinical trials at the new Alpha Stem Cell Clinics:

Sickle Cell Disease, Severe Combined Immunodeficiency, Beta Thalassemia, Lymphoma Cancer, Brain Cancer, HIV-AIDS, Lung Cancer, Kidney Failure, Stroke, Brain Injury, Leukemia, Retinitis Pigmentosa, and Spinal Cord Injury.[6]

There are also free conferences, where you can meet some of the champion scientists and patient advocates, and become a part of the battle for cure.

For instance, April 18, 2019, at the UCSF Mission Bay Conference Center in San Francisco, anyone who attended got to meet a veritable Who's Who among regenerative medicine, especially in the fight against sickle cell anemia.

Top scientists like: Arnold Kriegstein, Mark Walters, Alexis Thompson, Fyodor Urmov (great speaker), Don Kohn (famous for his work to save children from Bubble Baby Syndrome), and Alex Marson challenged our minds.

Sickle Cell specialists like hematologists Mehrdad Abedi and Alexis Thompson, social activists like Marsha Treadwell, patient care experts like Marci Moriarity, and inspiring personal stories from people like Cierra Danielle Jackson (who received — and deserved — a standing ovation).

Cancer specialists like Elizabeth Budde suggested a different method of treating cancer, using the immune system, Stephanie Jackson spoke about the practical details of caring for patients and families in a gene therapy trial.

---

[5] https://www.cirm.ca.gov/patients/alpha-clinics-network/about
[6] https://www.cirm.ca.gov/patients/alpha-clinics-network/alpha-clinics-trials?page=1

Stephanie Cherqui shared her work on a hereditary kidney disorder, ably assisted by patient advocate Shannon Keizer.

The crucial access to treatment? Danielle Bota spoke on that.

The National Institutes for Health was represented by Ed Benz, on Curative Therapies.

Ke Liu spoke on regulation and the federal Office of Cell Therapy.

Geoff Lomax spoke, whose business it is to know everything about the CIRM Alpha Clinics Network:[7]

"The clinics conduct FDA-authorized...clinical trials...aiming to achieve greater and more efficient results than the member organizations could if they acted independently. Each of the Alpha clinics has formed teams with specialized knowledge and experience....the network...has supported 60 clinical trials since 2015..."[8]

Another viewpoint on what the Alpha Clinics mean to the State of California?

"The Alpha Clinics are an integrated network of medical facilities that carry out clinical trials that have been given the green light by the Food and Drug Administration (FDA). They are state-of-the-art clinics that have expert physicians and highly trained nurses who are all experienced in the delivery of safe medicines to patients. Because these clinics operate as a network rather than as stand-alone facilities, they share best practices and collaborate on developing new ways of delivering stem cell therapies to patients. This is a uniquely California network and delivers a unique benefit to the people of the state."

Abla A. Creasey, PhD, Vice President, Therapeutics & Strategic Infrastructure.

Like CIRM? Subscribe to their outstanding weblog.[9]

The Alpha Stem Cell Network will be coordinated by a Stem Cell Center in San Diego. For more information, email them at: StemCellCenter@ quintilesims.com.

---

[7] https://www.cirm.ca.gov/events/4th-annual-cirm-alpha-stem-cell-clinics-network-symposium

[8] file:///C:/Users/Don/Downloads/Bioprocess%20ASCC%20Article%209_2018.pdf

[9] https://www.cirm.ca.gov/about-cirm/e-newsletters

7 "Tesi": Or, How to Engineer an Intestine

Dr. Tracy Grikscheit seeks to improve a defective intestine — to save an infant's life. (Huff Post Photo.)

Dr. Tracy Grikscheit's work involves something the world seldom thinks about — how to improve a defective intestine.

Why would we want such specific knowledge?

To save a baby's life.

First, consider the problem.

A premature infant may weigh as little as 500 grams. Born too soon, their intestines may be susceptible to a condition called Short Bowel Syndrome (SBS). "Portions of the small intestine (may be) missing or damaged at birth."[1]

At present, options for remediation are few, and none very good.

To save the child's life, a surgeon like Dr. Grikscheit, (M.D., at Children's Hospital Los Angeles) may perform a demanding operation. Like a length of rope, the intestines are lifted out of the baby's body — both ends still attached — and the non-working bits are removed. The remainder is sewn back together, and replaced.

---

[1] https://www.mayoclinic.org/diseases-conditions/short-bowel-syndrome/symptoms-causes/syc-20355091

What remains may not be enough to do its natural job. If the intestine does not have enough internal surface area to absorb nutrition, the child may need intravenous feeding, with complications of its own.

The child may require a bowel transplant: someone else's intestines are put in, with all the problems of donor availability, and the risk of the body rejecting it.

And afterward? A reduced intestine may not retain the food long enough to process it properly; the body passes it too quickly, without absorbing the needed nutrition. The baby may need to be fed intravenously. All too often, the baby's chances are slim.

As Dr. Grikscheit said:

"Small bowel transplants…have many problems, including poor graft survival, rejection, limited donor supply, surgical morbidity, and a lifelong (need for) immunosuppression."[2]

Instead — imagine the doctor saying: "We're going to make a TESI for your child!"

Tessy? What's that?

Making a usable and reliable Tissue Engineered Small Intestine (TESI) is one of the goals of Dr. Grikscheit's professional life.

A tiny dissolvable scaffold is made. It is hoped some of the baby's own cells will be put on this scaffold, possibly encouraged by a growth factor. This living tissue would be placed inside the baby. His/her life would go on, *and become normal.*

Is this possible?

Working on a grant from the California stem cell program, Dr. G. reports:

"We have shown that TESI forms when autologous cells (from the patient) are implanted on a polymer scaffold, and that TESI (duplicates the original) intestine … Importantly, animals recover from massive small bowel resection with TESI. Other regions of the gut (may also be made) via this approach.

"Our goal is to translate TESI…to an autologous human cell based therapy. Patients who needed emergency surgery for their intestine… could potentially be treated with TESI … Usable engineered intestines

---

[2] https://www.cirm.ca.gov/our-progress/awards/generation-and-expansion-tissue-engineered-small-intestine-human-stem-progenitor

would be far less expensive, more durable, and require less maintenance than any current therapy."

And a family goes home happy from the hospital...

As Dr. Grikscheit puts it: "We need to give these children a better future, measured not in months, but in decades like the rest of us."

8 Fighting Rett Syndrome

Eve Yi Sun began as a platform diver, and then shifted to a higher calling. (Personal photo.)

At the top of the ten meter platform, the young Chinese woman turned around, so only her toes gripped the board, and her heels hung suspended over the blue square of water far below. She raised her arms, poised for a second: two, three...

Flexing quadriceps and calves, she bounced higher and higher — and went for it, leaping, into the sky, arching, arms spread, defying gravity for one eternal instant, then tucking and rolling downward, faster, faster — straightening at the last, knifing through the surface...

Not many can dive, this ultimate physical expression.

But fewer still have the second skillset of Eve Yi Sun, PhD.

For years, Dr. Sun has used science to fight Rett Syndrome (RTT), a terrible disease which attacks young girls.

Perhaps one in ten thousand babies is born with RTT. At first, nothing seems wrong. The child moves and talks and breathes normally.

But when the child reaches 12–18 months, progress turns backward. Skills previously mastered are taken away: first she could talk — now she cannot. Once, she could move her hands purposefully, reaching and grasping; now that ability is gone, replaced by an endless hand-wringing.

The child may cry for hours at a time, inconsolably, as if mourning a normal life.

Boys rarely get the disease. If they do, it is typically fatal, before the age of two.

But it is the girls who suffer long: wheelchair lives of sixty or seventy years. Oddly, the ability to speak may return in later years, but there is no cure: not yet.

And that "not yet" is the reason I am glad Dr. Sun's career did not begin and end with springboard diving.

She works at two laboratories, one in Shanghai, China, the other at the University of California at Los Angeles.

The disease is caused by mutations in one gene: MeCP2.

According to the International Review of Neurobiology (in a chapter co-authored by Dr Sun):

"...Mouse genetics studies have demonstrated that the lack of functional MeCP2 in the central nervous system leads to RTT-like symptoms, <u>which could be reversed</u> (emphasis added — DR) upon MeCP2 restoration."[1]

"...could be reversed..." What a beautiful phrase!

Which brings us to three questions:

1. Without using people, how could doctors test for improvement, or even study how the condition begins?
2. Might regenerative strategies be useful?
3. How will the research be paid for?

---

[1] https://www.sciencedirect.com/science/article/pii/S0074774209890077

First, in research intended to bring cure, scientists need a *model*: like the previously mentioned mouse studies. But mice (though useful) are very different from people. Rats are larger and biologically closer to humans, but still there is a gap.

To answer that difficulty, Dr. Sun has developed a way to give RTT to non-human primates, a lab monkey species called cynomulgus.

Second, she has worked to use stem cells taken from the patient's own skin to develop cell lines of the disease, an inexhaustible supply of the "disease in a dish."

And third, most importantly (without it, all research stops), Yi Sun has received research funding from several sources, including the National Institutes of Health (4 grants of roughly $375,000 each),[2] the California Institute for Regenerative Medicine (CIRM) (a Basic Biology III grant for $1,382,400), as well as support from her lab in Shanghai.

Unfortunately, NIH grants may not be safe politically (cuts of 20% were recently proposed by the Trump Administration) and CIRM is running out of money.

Rett's Disease families are fighting back.

Rettsyndrome.org is an award-winning charity which has raised more than $46 million in the fight for cure.

One beautiful effort was the cross-country cycling ride of KC Beyers, stepfather of a daughter with Rett Syndrome. Imagine bicycling 2,600 miles! But he did it, and raised roughly a quarter of a million by that effort. — Tim Frank, Rettsyndrome.org

By every effort, small grants and large, the battle must go on, until the cure is found. It is not right that children and their families should go through so much.

We must end their suffering, find a cure for Rett's Syndrome.

And when that day comes, part of the credit will belong to Eve Yi Sun, PhD. — the woman who leaped into the sky.[3]

---

[2] grantome.com/grant/NIH/R01-MH082068-01A2
[3] http://tinyurl.com/y5mn4uch

9 The Disease Which Caused a Revolution?

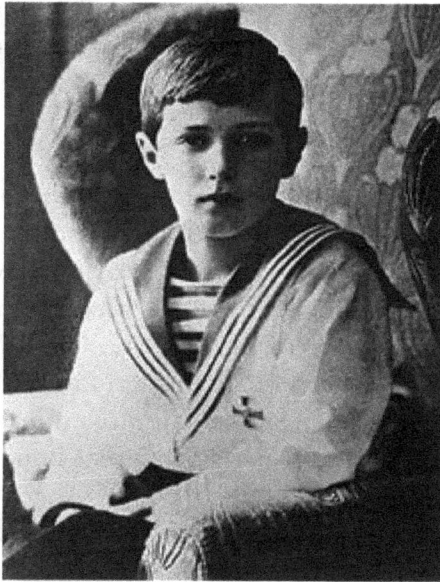

Tsarevitch Alexei, suffering from the "bleeder's disease," was taken off all medications by religious fanatic Grigori Rasputin…. (Wikipedia photo.)

It is said the Russian Revolution of 1917 might not have happened, had there been a cure for hemophilia, the "bleeder's disease." Untreated, hemophiliacs can suffer from internal bleeding in the joints, non-healing bruises, strokes from brain-bleeds, even death from blood loss.[1]

Tsarevich Alexei, young heir to the throne, inherited the disease from his mother, Empress Alexandra, who had gotten it by marrying a relative

---

[1] https://www.cdc.gov/ncbddd/hemophilia/facts.html

of Queen Victoria of England. Victoria shared the disease with many descendants, so that it became known as the Royalty disease.

The Russian royal family was consumed with the attempts to save their boy. which allowed the rise of religious fanatic Grigori Rasputin, who said he could cure the child by the power of prayer.

Despite a colossal ignorance of science and medicine, Rasputin did do one thing right. When he took the boy off all his medications, that included aspirin, now known to be a blood-thinner, deadly for someone with hemophilia.[2]

Rasputin gained much advisory power, including persuading the Tsar to take personal charge of his army in a war. Rasputin was reportedly a sex maniac and lived a life unrestrained by moral codes, at one point raping a nun.

Eventually, assassination was ordered.[3] Rasputin was fed poison, which did not work. A woman stabbed him in the stomach, but he survived. Ground glass was put in his meals, with no apparent result. Finally he was shot three times (once in the head), wrapped in chains, sewn into a heavy burlap bag and thrown into an icy river. When he was found later, one hand had burst through the sacking, and was sunk in the mud, seemingly dragging him ashore.

Rasputin's rise to power destroyed the reputation of the Royal Court, distracted the leadership from any serious attempt to solve the nation's problems — and by so doing perhaps helped bring about the 1917 revolution.

Today, perhaps 20,000 Americans have hemophilia A. Because women can carry the disease, but do not themselves usually suffer from it, most patients are male. Treatments may sustain life, but there is no cure.[4]

Working on a grant from the California stem cell program, Dr. Marcus Muench of the University of California at San Francisco (UCSF) is studying ways to challenge hemophilia.[5]

---

[2] https://www.amazon.com/Nicholas-Interrupted-Helene-Carrere-DEncausse/dp/0841913978

[3] https://en.wikipedia.org/wiki/Grigori_Rasputin#Assassination_attempt

[4] https://www.hemophiliafed.org/understanding-bleeding-disorders/what-is-hemophilia/hemophilia-a/

[5] https://www.cirm.ca.gov/our-progress/people/marcus-muench

He is not always easy to understand. When I interviewed him on the phone, I kept asking him to slow down, and he said: "I know, I know, lots of big words!"

His goal for hemophilia?

"... to test (if) ... growth factors such as VEG-F can improve engraftment of ... cells ... as a therapeutic approach for hemophilia A."

"My laboratory studies liver cells to better understand the ... immune system."

"Development of a cell therapy to treat hemophilia A ... may provide a long-lasting therapy or even cure for the disease, greatly impacting the lives of the patients and the economic burden that the disease places on them and the medical system."

Then came something which seemed important not just for blood diseases, but perhaps for other conditions as well:

"A driving force behind my basic research efforts is a desire for the research to be clinically applicable ... (with) the potential of (easing) or curing birth defects.

"... the current practice of ... transplantation is not yet effective enough to treat most (blood or liver) birth defects. My lab is using mouse models of ... transplantation to study methods of improving ... prenatal transplantation."

Maybe hemophilia could be prevented before the baby is born?

One of the causes of hemophilia is not enough of a growth factor, called Factor VIII.[6]

This might be made from skin cells. But one huge problem is that the new cells may not stick where they are sent.

The liver has little caves called sinusoids. What if those could be made larger and more sheltering for the new stem cells, for ease of engrafting?

The benefit they bring may be long-lasting, even permanent.

Because every child should be a prince or princess to their families.

---

[6] https://www.hemophilia.org/Bleeding-Disorders/Types-of-Bleeding-Disorders/Hemophilia-A

10 Leader of the Board

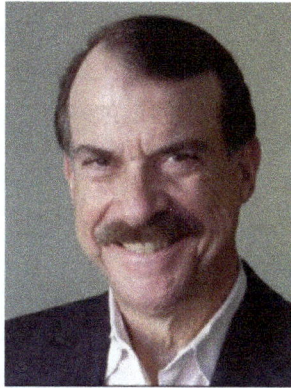

Jonathan Thomas had the unenviable task of following Bob Klein as chair of the board of directors — but built his own legacy. (CIRM photo.)

When you come to one of the board meetings of the California stem cell program, you will see up to twenty-nine people sitting behind long tables, arranged in a square. It is nothing fancy, no luxury: just each person has a pitcher of water and a glass before him or her, and a folded piece of cardboard with their name.

Here, at 1999 Harrison Street, in Oakland, California, decisions are made to ease suffering, save lives, and change the world.

As a citizens' initiative, Proposition 71, The California Stem Cells for Research and Cures Act, was approved by the voters and written into the State Constitution. The plans for the program were drawn up by Bob Klein, with the assistance of great legal minds like James Harrison. Klein was the first Chairperson, nominated into that position by the Governor, Lieutenant Governor, Treasurer and State Controller, and voted into that position by the entire board of directors.

Mountains of work lay ahead: to develop, fine-tune and implement the initiative.

The 29-member board of directors is a cross-section of the interested public. Technically titled the Independent Citizens Oversight Committee (ICOC), they are patient advocates from various disease groups, leaders from the biomedical field, expert researchers — appointed by numerous public officials.

https://www.cirm.ca.gov/board-and-meetings/board

I wondered if they would be able to work together, considering the authority positions they all held. Everyone was "the Boss" where they worked.

Perhaps realizing this was bigger than anything they would ever do — the board members left their egos at the door. They worked together, each expressing opinions fully, but always trying to move science forward, to benefit the patients.

Once I had a major disagreement with one board member. "Let's go out in the hall," he said, and we did.

We snarled and snapped at each other for a while, then agreed to disagree, shook hands, and went back into the meeting room, ready to work together again.

I never cease to be amazed by the excellence of those involved.

On the sides of the room sit 8–10 members of the hard-working staff: whoever is connected with the topics on the agenda. These are drawn from just 50 people who make the program come to life: California owes them a debt of gratitude.

Below is a brief interview with the Chairman of the Board, Jonathan Thomas. If I had to characterize him in one word, it would be "endurance". Like a long-distance runner, he is always on the move, doing something to benefit CIRM, and the patients the program must serve.

Chairman of the Board: **Jonathan Thomas**

How did you become chairperson?

In June, 2011, Bob Klein, founder of CIRM, was termed out. California Treasurer Bill Lockyer and Governor Jerry Brown nominated me to serve as Chairman of the governing board, the Independent Citizen's Oversight Committee (ICOC). After a process of interviews with officers of the board and patient advocates, my nomination was put to a vote of the ICOC, and I was elected.

How did it feel stepping into Bob Klein's position?

Overwhelming! I had (and continue to have) tremendous respect for Bob Klein and his creativity; his ability to design a new approach to medical research, building a program like nothing that had ever been before. CIRM was enabled through his vision. When the history of the regenerative medicine era is written, decades from now, Bob Klein will have a very prominent place. I have studied funding for many years, and CIRM is unquestionably "best in class". I was honored and delighted to have the opportunity to take what he designed, established and led — and to move it forward to another stage.

You had 15 years background as a patient advocate?

In the early 80's, I was asked to join the board of what was then the Crippled Children's Society of Southern California, now called Ability First. It is an organization that serves mentally and physically disabled kids. It has a number of centers, summer camps, and a variety of programs, including some which help find employment for challenged adults. I was Chair from 1990–1994, and a member much longer. The agency serves kids and young adults with disease and disability conditions like those which CIRM fights.

How did you become interested in biomed?

Biology was my principle interest as a kid; I always loved animals. In college, I had a double major: biology and history. Over time, I became very interested in molecular biology, which features critically in regenerative medicine. Becoming associated with CIRM was a great chance to revisit my core academic interests.

How did you first get involved in stem cell research?

I was working in finance, a private equity firm with no life science connections. But I had an interest in the possibilities of regenerative medicine. In 1998, Wisconsin scientist Jamie Thompson had a watershed moment in research, when he isolated an embryonic stem cell for the first time. I was looking for an opportunity to become involved in that promising new field. I heard about a scientist named Mike West, and his company: Advanced Cell Technologies. At this point, the whole notion of a stem cell company was at its earliest stage, and to get involved required real foresight from investors. It was not easy to get them to invest in the little-known science. But Mike West needed to raise early capital for his company. I undertook the effort, and helped him raise one million dollars.

You helped connect Los Angeles to the sea?

In 1994 I was appointed by Los Angeles Mayor Richard Riordan to serve on the Harbor Commission. This was a citizens' commission, planning a rail system from a dispatching yard in downtown Los Angeles to a port in the harbor. Called the Alameda Corridor, it was designed to increase the speed of delivery from ships to the city. Unfortunately, it had a 2.4 billion dollar funding gap, and no finance person. By then I had worked as a public finance investment banker, from 1985 to the early 2000s, so I had some experience with bonds. The project took several years, but we were able to fill the funding gap. The Alameda Corridor has been in existence almost twenty years now, doing precisely what it was intended to do. The plan required a variety of bonds and loans — and we finished it on the last business day of President Bill Clinton's term in office, signing it into law in the West Wing.

As a lawyer, I clerked for the DC circuit, the federal appellate court in DC, viewed as the 2nd highest court before the Supreme Court. I had the privilege of working in the White House council office in the last summer of the Carter Administration, and also worked as a litigation lawyer in Los Angeles. Working at CIRM checked a lot of boxes for me: science, public finance, government, law, and patient advocacy.

Fundraising experience — with lions and elephants?

Before joining CIRM, I was on the board of the Los Angeles Zoo for many years. There was a constant need for funding, and I was asked by the President of the City Council to make a plan to help get a bond measure through. The goal was to provide funding for construction of major animal exhibits, delayed for lack of funding. I helped put together a campaign, and we raised $50 million to fund zoo projects, including an elephant enclosure, a reptile house and a tropical rain forest. I also worked on a private equity firm, which required me to raise funds from institutional investors. For various projects, we raised approximately $200 million.

How would you describe your job at CIRM?

I oversee CIRM's governing board, the Independent Citizens Oversight Committee (ICOC), and every endeavor the board has a hand in. This encompasses a great many things, including the science itself, financial and government issues, legal matters, communication, and others. The chair is involved in looking for funding, as well as performing an ambassadorial role, dealing with major stakeholder groups, developing a network of relationships, knowing the major players in healthcare and business, being aware of grantee matchups with industry. There is also a "Sacramento element". We need to stay in touch with our governmental leadership in the Capitol. Here we are blessed with the leadership of Senator Art Torres (Ret), our Vice Chair. We benefit greatly from his knowledge and experience. The Chair's job also includes major communications functions — cooperating with reporters and columnists, editorial boards of newspapers, and the Internet. No two weeks are the same, and there is never a shortage of things to do. Fortunately we have an outstanding team, which makes everything possible.

You took action to eliminate the perception of conflict of interest on the board — describe what you did, and why?

Since its inception, the ICOC has always had 13 members from institutions eligible to receive grants. Over time, there have been criticisms that this might create a conflict of interest, that some might use their position to get grants for their parent institutions. There have not been any instances of that, but the possibility has been raised. This came to a head in December 2012, when the Institute of Medicine (IOM) raised the issue as a concern. It was my view that we needed to address this criticism once and for all. We spent a solid month developing a response to all criticisms, and in January 2013 the Board approved a comprehensive position on conflict of interest. The 13 members representing institutions eligible for funding would no longer be allowed to vote. They could participate, and share knowledge, but not vote. It was drastic action, but in my view, it was needed, or this issue would be endlessly raised, with or without justification. I believe this erased even the perception of conflict of interest. When we instituted the corrective measure, reaction was positive and immediate, from the Institute of Medicine and others.

(Note: This was a bold decision, an intelligent attempt to solve a real problem. But some feel the solution goes too far, and may have a chilling effect on board participation. Personally, I am concerned that an over-restrictive policy denies us the involvement of the experts on the board. I prefer that everyone should be allowed to contribute freely, just not allowed to vote on any grants which could go to their parent institution. I also feel that opponents of the program cannot be placated by any policy; they are against government programs in general, and will oppose us no matter what — DR).

Can you name a favorite grant or two?

That is a difficult question. I am exceptionally impressed with all the work we fund, and would match the quality of our scientists' work against any in the world.

With that stipulation noted, Don Kohn's UCLA work on Severe Combined Immune Disorder (SCID, "Bubble Baby" disease), is a great example of the interface of stem cell and gene therapy. The clinical trials have cured more than 50 children, and that — helping patients and their families — is what CIRM is all about.

Another spectacular success was made by Dr. Tippie Mackenzie and her work at the University of California at San Francisco (UCSF). Alpha Thalassemia is a deadly blood disorder. To fight it, Dr. Mackenzie takes stem cells from the mom's bone marrow, injects it into the unborn fetus, correcting the condition — before the baby is born.

After the grant is awarded, does CIRM provide other help to scientists?

CIRM is unique as a funding agency, in its ongoing support for grant recipients. Unlike some groups, who pat you on the back, wish you good luck and then leave you all on your own, CIRM is an active participant throughout the entire life of the grant. Advisors check up on the scientists and help him/her deal with any problems that may arise, including interactions with the FDA.

If California Renews CIRM, What Good May Come?

A tremendous amount. First, a lot of our projects have a way to go before proof of concept is established. This means the big money folks are generally not involved, holding back until after phase 2 or phase 3 of the clinical trials are completed. As this takes 6–10 years, much of CIRM science is not at that stage yet; we need to provide funds to bring them forward, through this time period. Secondly, we need to provide funding for the next generation of scientists, those who will build on the work pioneered by Prop 71. This is a new field, much of it built on California's contribution. If CIRM can continue to accelerate the field, it would be a huge benefit for California, the nation and indeed the world.

11 Hitting Yourself in the Chest

Eloquent speaker for cystic fibrosis research, Siri Vaeth takes joy in the singing advocacy of her daughter Tessa Dunn. (Huff Post photo.)

When our children Desiree and Roman were young, the four of us would climb into bed on a Sunday morning and read the funny papers.

One day Gloria had a cold, and started coughing. Roman said to her, "Doctors say you need to hit yourself in the chest, to break loose the congestion."

He said it so reasonably, I almost believed it — and his Mother did.

"Like this?" she said, thumping herself carefully.

"Oh, no," said Roman, "You have to hit much harder than that!"

I forget who broke and laughed first, but it became a family classic, when Roman conned Mom into hitting herself.

But there is a condition which requires you to do just that.

For people who have cystic fibrosis (CF), the chest builds up heavy mucus inside, so you must literally hit yourself every morning: pounding your ribs to loosen the phlegm and cough it up. And if you cannot do it for yourself, either parents or a machine (an electric vest!) must do it for you.

Cystic fibrosis is a lung disease so severe your only hope of extending life may be a double lung transplant — a dangerous operation you may not survive — and that is only if someone's loved one has recently died and the family is willing to donate.

But CF is more than that. It impacts nearly every organ system in the body.

There are digestive difficulties: your body can't absorb and retain enough nutrients, so that you are starving even while you eat. Children in school may require constant trips to the bathroom.

There is major pain — in the sinuses, chest and joints.

Reproductive problems: in men, the vas deferens (connection between penis and testicles) may fail to form, so the sperm cannot travel. Women may also have challenges conceiving, due to thick cervical mucus.

Osteoporosis (thinning of the bones) is common, due to poor absorption of vitamins and minerals. And it is likely you will face repeated sinus surgeries, due to the regrowth of polyps.

You will need respiratory therapy, two to six hours every day — you may get diabetes as a side effect — and you will almost certainly die young.

In 2003, during the struggle to pass California's stem cell initiative, Proposition 71, I heard an excellent speaker talk about CF. She made the nightmare condition seem real, as if it was happening to me, and also that something must be done about it.

Not long ago, when I was looking up information about CF, I ran across an unusual name, Siri Vaeth. It sounded familiar, so I Googled her, and sure enough, there she was, the excellent speaker!

Executive Director for an organization called Cystic Fibrosis Research, Inc. (CFRI), Ms. Vaeth is doing what she was doing then, advocating for patients.[1]

---

[1] http://tinyurl.com/yyl6eknk

Her fight is deeply personal. Here's why:

"My daughter Tess was diagnosed with cystic fibrosis in 1995 when she was five months old. In addition to her extreme failure to thrive, she developed pneumonia. I have feared for her survival ever since...

"She has had multiple lengthy hospitalizations...caused by damaging lung infections, as well as five plus surgeries, and countless multi-week intravenous antibiotics treatments.

"She takes nearly 50 pills a day, injects insulin to manage CF-related diabetes...and spends a minimum of two hours a day doing respiratory therapy.

"But she is a warrior, powering her way through the physical challenges to pursue her life goals."

As for her organization, Cystic Fibrosis Research Inc., Siri says:

"CFRI has been a tremendous resource and...community for me since Tess' diagnosis...We provide programs to the nationwide CF community. We fund innovative research, we advocate on...the state and federal level, provide educational conferences, podcasts and printed resources... provide counseling, support groups, stress reduction classes — we are small but mighty!"

Advocates need scientists, and scientists need funding, whereby we work together.

One scientist working on cystic fibrosis is Stanford's Matthew Porteus, and some of his research is funded by CIRM.

"Cystic Fibrosis (CF) is one of the most common genetic diseases in California. There is no curative therapy for CF, and patients spend a lifetime focusing on mitigating the symptoms...Moreover, costs of treating (even) a single CF patient are enormous..." — Matthew Porteus.[2]

Porteus is attempting to correct the genes that have gone wrong in CF patients: first for the one symptom, sinusitis, and then for a "Universal... correction for the other CF-causing mutations."

It is crucial that such research go forward.

Unfortunately, Washington may let us down.

Consider the "Orphan Drug Act" (ODA), which conservatives may repeal.

For thirty years, Congress has offered a 50% tax exemption for research on cruel diseases like CF, also allowing the researchers to

---

[2] https://www.cirm.ca.gov/our-progress/awards/genome-editing-correct-cystic-fibrosis-mutations-airway-stem-cells

have exclusive rights to products developed for seven years. This is an incentive to develop a cure for a condition which does not have the numbers for mass market sales, an "orphan disease" which might otherwise be ignored.

Now, that incentive for progress may be taken away.

"The repeal of the orphan disease tax credit would be a major blow to the furtherance of treatments and cures for the rare disease community...Without new treatments the taxpayer will be putting out an immeasurable amount more in terms of hospitalizations, surgeries, and other needed treatments..." — Jacob Fraker, CFRI Legislative and Government Affairs Analyst.

Fortunately, there are still friends of research on both sides of the aisle, and as of this moment the Orphan Drug Act still exists in law.[3]

For more information, contact the URL below.[4]

But for right now, *listen to a beautiful song*, "Breathe", which should be the national anthem for all who suffer an orphan disease.

One of the people singing is Tessa Dunn, singer, songwriter, and the daughter of Siri Vaeth.[5]

---

[3] https://rarediseases.org/orphan-drug-act-resolution-introduced-in-the-house-of-representatives/

[4] http://rareaction.org/saveorphandrugs/

[5] https://www.engagecf.org/news/2017/11/8/cfri-many-voices-one-voice-cf-advocacy-awareness

12 The Cost of Doing Nothing

Some say: "We cannot afford to pay for stem cell research — it is too expensive!"

I feel just the opposite — we cannot afford NOT to pay for regenerative medicine.

Before we consider the cost of doing nothing, just trying to live with chronic disease, consider the expense of renewing the California Institute for Regenerative Medicine (CIRM).

We are asking $5.5 billion. Tax-free state bonds will be sold to raise the money, just as was done for the first CIRM.

How will the funding of CIRM affect you, in your personal finances?

It won't. The payments will come from existing state revenues. You did not feel any economic anguish from the first CIRM, did you? I did not, myself.

How much will CIRM cost the state?

If you average the cost of the bonds across their 40 year term of repayment, even accounting for interest, it works out to $4.89 cents a year, per person, per year.

Less than five dollars per person for an all-out attack on disease which costs our country $3 trillion dollars a year?

It would be the greatest bargain in medical history.

And the costs of doing nothing? That is a recipe for bankruptcy. Look at just some of the expenses involved:

First, one gigantic number — **$3 trillion dollars**. That is one year's cost (rounded off from $2.97 trillion) of chronic (incurable) disease in America, April, 2019.[1]

---

[1] https://www.cdc.gov/chronicdisease/about/costs/index.htm

How much is $3 trillion? As much as America's individual federal income taxes ($1.8 trillion) and payroll taxes ($1.2 trillion) — combined.

Am I making this up? My source for the government figures is the Congressional Budget Office (CBO).[2]

Let's break it down.

The Federal Center for Disease Control and Prevention page gives cost statistics for several major diseases:[3]

Cancer — $174 billion
Diabetes — $237 billion
Obesity — $147 billion
Arthritis — $140 billion
Alzheimer's disease — $159 billion

The next three pages you don't really have to read; just glance at them and if something strikes you as interesting, the URL for more info is there.

They show some of the expenses we never think of, till they happen to us.

In another chapter later on, we'll see how medical costs are putting more and more people into bankruptcy.

But for right now, glance quickly at some of the unexpected expenses of chronic disease... This next section is confusing, but amazing...

Bone Marrow transplant operation: $665,000.
https://www.technologyreview.com/s/602113/gene-therapy-cure-has-money-back-guarantee/

Hospital costs for the loss of a child who dies of infection: $300,000. This one really infuriates me. Even if the child dies, still the hospital bills must be paid.
https://www.ncbi.nlm.nih.gov/pmc/articles/PMC4846488/

Fetal alpha thalassemia: (1985 estimate lifetime costs): $1,342,140.
https://ajph.aphapublications.org/doi/pdfplus/10.2105/AJPH.75.7.732

Sickle cell disease: (2004) $488 million annual US cost. http://www.bloodjournal.org/content/124/21/5971?sso-checked=true

Acute Myeloid Leukemia: roughly $50,000 for one treatment. https://www.ncbi.nlm.nih.gov/pubmed/11276372

---

[2] https://www.cbo.gov/topics/taxes
[3] https://www.cdc.gov/chronicdisease/about/costs/index.htm

Bone Disorders: osteonecrosis: national cost estimated at $1.6 billion annually. https://www.medscape.com/viewarticle/452499

Osteoarthritis: in 2013, $16.5 billion in hospitalization, 4.3% of all hospitalizations.

https://www.cdc.gov/arthritis/data_statistics/cost.htm

Diabetes: $245 billion annually. https://www.cdc.gov/diabetes/data/statistics-report/deaths-cost.html

Macular degeneration: nearly $10 billion annually. https://iovs.arvojournals.org/article.aspx?articleid=2414895

Retinitis Pigmentosa: $7,317 per person per year. https://www.researchgate.net/publication/225086077_Health_Services_Utilization_and_Cost_of_Retinitis_Pigmentosa

Heart disease in all forms costs America $316 billion in both health care costs and lost productivity: this represents 1/6 of every health care dollar.

https://millionhearts.hhs.gov/learn-prevent/cost-consequences.html

HIV/AIDS: the most recent lifetime cost estimate for an individual's HIV treatment was $379,668.

https://www.cdc.gov/hiv/programresources/guidance/costeffectiveness/index.html

Kidney Disease: In 2016, treating people with End Stage Renal Disorder (ESRD) cost $35 billion. https://www.cdc.gov/kidneydisease/basics.html

Amyotrophic Lateral Sclerosis: $256 million annually. https://alsnewstoday.com/2017/06/08/cost-neuromuscular-disorders-u-s/

Spinal Cord Injury, individually: high tetraplegia first year $1,000,000, then $184,000 annually; paraplegia starts at about 518,000 first year, then about $69,000 annually. https://www.spinalcord.com/blog/what-is-the-real-spinal-cord-injury-cost

Stroke: $34 billion annually. https://www.cdc.gov/stroke/facts.htm

Solid Cancers: $157 billion annually. https://www.ncbi.nlm.nih.gov/pmc/articles/PMC3107566/

Total annual cost of cancer care in U.S. is projected to reach $175 billion by 2020.

https://www.cancerinsurance.com/blog/ten-statistics-on-the-cost-of-cancer-treatment-in-america (American Society of Clinical Oncology)

Blindness, Vision Loss: $139 billion annual cost. https://www.preventblindness.org/sites/default/files/national/documents/Economic%20Burden%20of%20Vision%20Final%20Report_130611_0.pdf

Glioblastoma: approximately $53,000 per individual annually. https://link.springer.com/article/10.1007/s40273-014-0198-y

Leukemia: acute leukemia, first year: $463,414. https://www.mdedge.com/hematology-oncology/article/184586/anemia/financial-burden-blood-cancers-us

Advanced Malignancies: $4 billion out of pocket for cancer patients; total national costs $87.8 billion. https://www.fightcancer.org/sites/default/files/Costs%20of%20Cancer%20-%20Final%20Web.pdf

Myelofibrosis: $54–68,000 average per year over three years. http://www.bloodjournal.org/content/120/21/972?sso-checked=true\

Polycythemia Vera and Essential Thrombocythemia: $29,553 annually. http://www.bloodjournal.org/content/120/21/2071?sso-checked=true

Brain Cancer: $150,000 annually. http://blog.braintumor.org/brain-tumor-facts-figures-march-2018-cost-of-care/

Ataxia: annual cost of one person's care for Friedreich's Ataxia is $24,859. https://www.eurekalert.org/pub_releases/2013-02/bc-wi022713.php

13 The ATM Disease

To save an unborn child's life, Tippi Mackenzie uses stem cells from pregnant mother's bone marrow. (Stem Cellar photo.)

You know those ATM machines in front of banks, where you can make deposits or withdrawals of money? Imagine an ATM which collected blood. You go there once a month, stick your arm in a slot — and a needle would stab into your vein?

There is a disease with certain similarities to that. It is called ATM (Alpha Thalassemia Major) and it requires transfusions of blood — every month.

If you had the life-threatening disease, you would need those transfusions. Twelve times a year you would set up a medical appointment, endure the stick of needles in your arm, pay the endless medical bills — and that is after you find a compatible blood supply!

There may be complications. Too many blood transfusions may concentrate iron in your body, building up deposits around heart, liver and lungs, interfering with their function; these must be removed by a process called iron chelation.

Some patients with the ATM disease last to the age of 20 or 30. For others, death may come in the womb itself.

ATM "is so severe that affected fetuses will develop anemia due to the lack of functioning red blood cells...(which) can lead to heart failure..."[1]

A disease which kills unborn children? And which may affect the mother as well? This must be fought.

With the aid of a CIRM grant ($12.1 million), Dr. Tippi Mackenzie, MD, is taking on this challenge.[2]

She has been preparing for this moment for many years. Not only is she an associate professor in Pediatric Surgery at the University of California at San Francisco (UCSF), but she is also co-Director of the UCSF Center for Maternal–Fetal Precision Medicine.

Her approach? Dr. Mackenzie intends to take stem cells from the bone marrow of the pregnant mother, and inject them into the fetus — while it is still in the womb.

The advantage is clear. The mother's stem cells will not be rejected by her child.

This will be a clinical trial: a safety test. The outcome is not clear. But the risks of doing nothing are plain. This is a nightmare condition.

ATM threatens millions. Once called the "Chinese disease" for its prevalence among Asians, ATM is "one of the most common genetic disorders worldwide, with prevalence the highest in China, South East Asia, Africa, Middle East, and India" — and those of Mediterranean descent — an astonishing number of people![3]

"The UCSF Benioff Children's Hospital Oakland and the Center for Maternal–Fetal Precision Medicine have begun enrollment for a clinical trial that will test the safety of combining in utero hematopoietic stem cell transplant with a fetal transfusion of red blood cells."

"This combination is aimed at treating, and possibly curing, Alpha Thalassemia Major (ATM) a blood disease that is often fatal in utero. This trial is the first of its kind in the world, and could also lead to treatments for other life-threatening blood diseases, such as sickle-cell anemia ..."[4]

---

[1] https://fetus.ucsf.edu/alpha-thalassemia

[2] http://www.cirm.ca.gov/about-cirm/newsroom/press-releases/06292017/stem-cell-agency-invests-more-44-million-treatments

[3] https://pedsurg.ucsf.edu/news--events/ucsf-news/72384/CIRM-Awards-Tippi-MacKenzie-MD-$121M-to-Treat-Fetuses-in-the-Womb-Alpha-Thalassemia-Major

[4] http://thalassemia.com/services-intrauterine-therapy.aspx#gsc.tab=0

But the procedure is no longer just a hopeful theory — ask the parents of Eliana Constantino, who received "... the first ever in utero stem cell transplant to treat alpha thalassemia"... Their little baby is just fine.[5]

Godspeed, Dr. MacKenzie: may your procedure save the lives of unborn millions.

---

[5] https://blog.cirm.ca.gov/tag/alpha-thalassemia/

14 Preventing Medical Bankruptcy?

Elizabeth Warren. Are people becoming more — or less — threatened by medical bankruptcy? (Wikipedia photo.)

My recent bout with cancer left a medical bill of $990,000 — almost a million dollars for surgery, radiation, hormone injections, etc., etc...

How could I possibly pay such a bill? I am a senior citizen (age 74) on a fixed income. I have no savings, zero stocks or bonds. The only asset of value Gloria and I possess is our (heavily mortgaged) house.

What could we could do? Declare bankruptcy? Sell the house and use the equity to leave the state, find a cheaper place to live? What would that mean to our paralyzed son, Roman Reed, for whom I provide unpaid attendant care?

Fortunately, our insurance (Kaiser) covered everything. My co-pay was zero.

Without insurance, I could not have afforded to fight the cancer. Instead, I am (according to the latest blood tests) fine and dandy.

Not everyone is so fortunate.

Bankruptcies for medical causes have been increasing for years. Compare:

1981: "Only 8% (eight) of families filing for bankruptcy did so in the aftermath of a serious medical problem..."

2001: "... illness or medical bills contributed to about <u>half</u> (emphasis added — DR) of bankruptcies..."

2007: "62.1% of all bankruptcies (in America) were medical ..."[1]

I am quoting a major study by Harvard Law School: "Medical Bankruptcy in the United States, 2007: Results of a National Study." One of the authors was Elizabeth Warren. I did not realize that before Senator Warren was a politician, she was a lawyer who studied medical bankruptcy.

That study was referred to by President Barack Obama in his 2009 State of the Union address. (2) Of medical bankruptcy, he said:

"We must address the crushing cost of health care. This is a cost that now causes a bankruptcy in America *every thirty seconds*. By the end of the year, it could cause 1.5 million Americans to lose their homes."[2]

I appreciate both Senator Warren and President Obama's work on Obamacare.

The Affordable Care Act (ACA, Obamacare) brought down medical bankruptcies, cutting that annual rate by half.

"Filings have dropped about 50%, from 1,536,799 in 2010 to 770,846 in 2016."[3]

But no matter where you stand on the issue of health insurance — whether you support a national program, a privately-paid program or something in between — we can all agree on one thing:

The best prescription to eliminate bankruptcy is cure.

To lower the cost of medical treatment (assuming abandonment is off the table, we are hopefully not going to let people just die) we must *cure the diseases.*

Consider polio.

If Jonas Salk had not invented the polio vaccine, America would now be paying billions of dollars a year just to keep people alive: not fixing

---

[1] http://www.pnhp.org/new_bankruptcy_study/Bankruptcy-2009.pdf
[2] https://obamawhitehouse.archives.gov/the-press-office/remarks-president-barack-obama-address-joint-session-congress
[3] https://www.consumerreports.org/personal-bankruptcy/how-the-aca-drove-down-personal-bankruptcy/

them, just maintaining them in iron lungs, lying there, immobile, suffering until they died.[4]

Instead, with rare exceptions, polio is gone. We do not have to pay those incredible bills, suffer the agony of living in iron lungs, or watch our loved ones slowly die.

The "March of Dimes", begun by President Franklin D. Roosevelt (who suffered a polio-like condition called Guillain–Barré Syndrome) and his friend Basil O'Connor, was patient advocacy in action.

On January 30th, 1938, President Franklin D. Roosevelt spoke of that glorious initiative, saying:

"Yesterday between forty and fifty thousand letters came to the mail room of the White House. Today an even greater number — how many I cannot tell you — for we can only estimate the actual count by counting the mail bags. In all the envelopes are dimes and quarters and even dollar bills — gifts from grown-ups and children — mostly from children who want to help other children get well."[5]

A President with a disability — and the children of America — united, unstoppable!

Making people well, instead of watching them go broke and die... is that not an example to emulate?

---

[4] http://www.euro.who.int/__data/assets/pdf_file/0005/345686/Case-study-US-polio.pdf?ua=1

[5] https://en.wikipedia.org/wiki/March_of_Dimes

15 New Babies, New Scientists

Mana Parast: "On an average day in California, 149 babies are born prematurely.... (UCSD photo.)

Forty-seven years ago, Gloria and I knew our unborn child would be a little girl, a "Baby Woman!" as Gary Trudeau put it in a famous Doonesbury cartoon.

I also knew what her name should be! In the eighth grade I saw a book, "*DESIREE,*" by Anne-Marie Selinko, about Napoleon Bonaparte's first love. I didn't read the book because I don't like romances — but I loved that name, Desiree, pronounced like treasure-ray...I saved it up in my mind, all those years.

"I have the perfect name!" I said.

"Oh no," said Gloria Jean, "I have her name! It has to be — "

"What?" I said, knowing she would win. She could have whatever she wanted right now, being "enceinte" with our baby. If she wanted to name our girl Fred, I knew I had no choice. I held my breath...

"I read a book," she said, "About Napoleon's first great love!"

Same book! Same name! Desiree!

But as the pregnancy progressed, I had thoughts I could not share. Would she be all right, our baby to be?

I wanted to be in the delivery room, when our first baby was to be born.

The hospital prepared us with a movie about birthing. It would be a little graphic, we were told. I figured I would have to comfort Gloria.

But when the movie started, and the blood began, I cringed back in my chair. "Oh, this is interesting!", said Gloria, leaning forward.

Every night we did the LaMaze breathing exercises, getting ready.

Back then, it was not usual for a husband to be in the birthing room, and the doctor had warned me I would be ejected if problems arose, i.e. if I fainted.

And a friend told me, that whatever I did, "Don't look at the 'after-birth', that I would be grossed out if I saw the placenta...

We were standing in line at the bank when Gloria's water broke. A small boy loudly told his mother about "the lady wetting her pants!", but Gloria is a determined person, and stayed long enough to complete her transactions.

We dashed home for her suitcase, and rushed to the hospital.

In the delivery room Gloria was thirsty and hungry, but was not allowed to eat or drink: just a few ice chips.

We waited. Gloria wanted her legs lifted, and I held them up as long as I could — then she told me to get down on all fours and let her rest her feet on my back, so I did that too, until her contractions came too rapidly, and it was time.

Gloria strained like a weightlifter, red-faced and yelling. I stood by her head, repeating the phrases we had agreed on, "choo-choo breath, choo-choo breath" (shifting the breathing to the upper chest allows birth muscles to work more efficiently), while the professionals did what professionals do.

I heard words I did not understand... "dilating centimeters... head beginning to crown..."

And then...

"Mr. Reed, you have a little girl — who just urinated on me!" said the doctor.

I could not help replying:

"That, doctor, is the beginning of a lifelong lack of respect for authority!"

Suddenly, I held my daughter in my hands, supporting her neck properly, what a priceless treasure. Some people say newborns are ugly, all scrunched up by the birthing process. Maybe theirs! But my daughter was perfect, in every way.

Desiree...She looked at me. And she smiled.

I watched her first moment in the world. She seemed an in-charge little being, very much not intimidated by the change in her environment, and the tremendous struggle she and her Mother had just been through.

Her gaze flicked to her tiny hand, She closed her little fingers, frowning slightly, like, what is this?

The nurse took her from me; I felt like growling.

And then I remembered, what was that thing I would be shocked by, the afterbirth, the placenta? I asked the doctor, who pointed.

And there it was: a little pile of white plastic-looking stuff, a few streaks of blood, nothing gross. What I felt was — a sense of gratitude. In my mind I said "thank you... for sheltering my baby all those months".

"You can give your wife this," said the nurse, handing me a bowl of green jell-o. But when I fed it to her, the gelatinous cubes fell out of her mouth, she was too tired to chew...

In the waiting room, I grabbed Gloria's mother, Soledad, and lifted her, threw her high in the air.

"My baby is born, my baby is born!" I shouted, while Gloria's mom screamed, "Put me down, are you crazy?!"

A seeming digression: remember the sword-fighting climax to Shakespeare's "MACBETH"? Two warriors, Macbeth and Macduff, are dueling, and Macbeth tells his rival he has magic and cannot be killed "by any man of woman born."

But Macduff just laughs and says:

"From my mother's womb, I was untimely ripped!"

"Then lay on, Macduff, and damned be he who first shall cry enough," shouts the Scottish lord. The swordfight continues, and Macbeth loses his head, literally.

We are not told, but apparently Macduff's mother gave birth to him prematurely.

And what is the most common cause of premature delivery?

A stem cell deficiency called pre-eclampsia.

"Pre-eclampsia is a pregnancy complication...(which) threatens 5–8% of all pregnancies. It has major effects on blood pressure and kidney function of the mother. It is responsible for a significant proportion _of maternal deaths and growth-restricted babies_..."[1]

I had not considered that. We had gone into the hospital to have a baby, and our thoughts were on the soon-to-be-born child — but Gloria had been at risk? And the same condition, pre-eclampsia, might have caused our child to be born early? We had been oblivious to the danger.

"On an average day in California, 149 babies are born prematurely. Many of these babies will require weeks of care in an...intensive care unit..."

Dr. Mara Parast, of the University of California at San Diego, is trying to reduce that suffering.

Consider the stem cell challenge.

In pre-eclampsia, the body makes too few "tropho-blasts", a stem cell that builds the placenta.

Dr. Parast has made a stem cell model of both the deficient trophoblasts and the healthy ones, to compare the differences, maybe use the healthy stem cells to replace the sick ones — and also test new medications.

"Parast's...lab has created a human trophoblast stem cell model — a first — that can be used to study stages of placental development. The CIRM grant will fund this work... to identify potential stem cell-based therapies for treating pre-eclampsia..."

But what if this outstanding young doctor had not been able to get a grant? The average age for first grants from the National Institutes of Health (NIH) nowadays is 44 years. There are so few grant opportunities it may take decades to find one — whereby some scientists, unfunded, have no choice but to leave the profession.

Now whenever I have a really heavy-duty question about CIRM, I try to ask Bob Klein, the man who began the program. This is not always

---

[1] https://www.cirm.ca.gov/our-progress/awards/human-pluripotent-stem-cell-based-therapeutics-preeclampsia

easy to make happen. Bob's schedule is slightly more crammed than the Pope's.

But I caught him in the hallway on the way back from the restroom, and asked: if he had to choose, which of CIRM's many grants was the absolute most important?[2]

He answered without a pause: "The grants that help young scientists to survive financially, while trying to find their place in the stem cell field…"

Later I asked another expert, Dr. Pat Olson, Chief Scientist of CIRM, if there were any grants like that, specifically targeting young scientists. She named four: New Faculty grants part I, which had been given to 22 investigators, part II of that grant went to 23 more, the Physician Scientist Translational Research Awards — another 15 — and the Medical College Repayment Award, provided to 5.

Think what that means: 65 scientists who might otherwise not been able to fund their research. They might have had to leave the field. Instead, they are working.

Dr. Arlene Chiu, former Chief Scientific Officer, adds this:

"These grants are designed to encourage (new) investigators to pursue bold and innovative studies…providing salary and research funding for up to five years, ensuring that they have stable, secure financial support as they begin their…scientific careers."[3]

New scientists need stability for fledgling careers — like new babies being born.

P.S. In later years, Desiree Reed went on to become the world's first female Hispanic Athletic Director. She is currently directing the sports program at the University of Nevada at Las Vegas (UNLV).

---

[2] https://health.ucsd.edu/news/releases/Pages/2012-12-12-parast-awarded-CIRM-grant.aspx

[3] https://www.cirm.ca.gov/about-cirm/newsroom/press-releases/06282007/stem-cell-institute-solicits-new-faculty-award-proposal

16 Secrets for Free

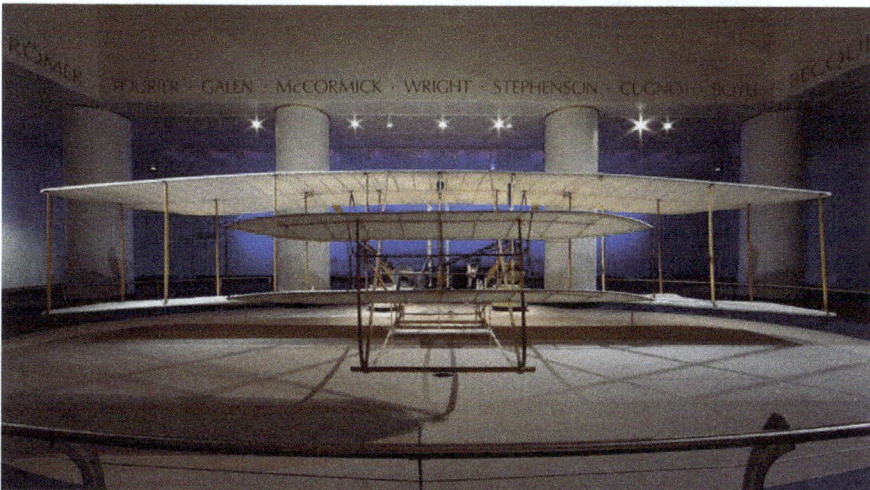

When the Wright brothers built their plane...they were not working alone.... (Museum of Science and Industry photo.)

If you visit the Chicago Air Museum (now called the Museum of Science and Industry) you can see the actual plane the Wright brothers flew, on the world's first motorized flight. It is an amazing museum, well worth the admission fee.[1]

One of the most interesting exhibits, though, is easy to miss: a bulletin board under glass — *letters written to and from* the brothers Wright.

When Orville and Wilbur were building their plane, getting ready to fly at Kitty Hawk, they were not working alone. Around the world, other

[1] https://www.tripadvisor.com/LocationPhotoDirectLink-g35805-d131645-i155108674-Museum_of_Science_and_Industry-Chicago_Illinois.html

researchers were trying to do the same thing — and sharing information on the subject.

Just as aeronautics engineers needed to know about propeller shapes and wind speed, tips they got from one another, even so do stem cell researchers learn from other scientists, in personal communication and published work.

That interchange is vital. If you are reading this book, you are probably good at info searching yourself. As you know, if you go to Google, type in "Google Alerts" in the search bar, and go where it sends you, it will tell you how to sign up for whatever your interest is. After that, your email will typically have four or five clickable links every day.

Unfortunately — some of the best info-sources are pay sites now. The Wall Street Journal does excellent articles, available online, but you have to pay. The New York Times and Washington Post also have pay sections, but they do allow you a certain number of free articles every month.

More and more information is only available for money.

I found one really interesting article about the growth of regenerative medicine — reportedly soon to be a $50 billion a year industry — but the article would cost $4,599 to read!

Forty-six hundred bucks for an article? I will pass on that, thank you.

Fortunately, if you look around, there are summaries available, and here is one.[2]

The California stem cell program has a completely different outlook.

CIRM exists to help California, our country and the world: partly by sharing information. California pays for that knowledge, but you may share, free.

For instance, as the father of a paralyzed young man, I am naturally interested in possible cures for paralysis. So to find out what's new, I go to www.cirm.ca.gov, type in "spinal cord injury" and I get this — a lot.[3]

If you go there, you will see (in addition to a picture of my handsome son Roman!) numerous videos and explanatory pieces — and links to many attempts to cure or ease paralysis. Some study the cure as a whole; others focus on one aspect of the problem — from general to very specific!

---

[2] http://tinyurl.com/yaqtt5d7
[3] https://www.cirm.ca.gov/our-progress/disease-information/spinal-cord-injury-fact-sheet

Below is a sample of the sort of stuff you will find: scientists, where they work, and a hint of their approach: it is a quick index of paralysis cure research, all neatly organized and ready to read, with links for in-depth study.

Some are mentioned twice because they are approaching the problem two ways.

Brian Cummings, University of California, Irvine: the Immunological Niche: Effect of...drugs on stem cell proliferation, gene expression, and differentiation in a model of spinal cord injury.

Mark Tuszynski, University of California, San Diego: Neural Stem Cell Relays for Severe Spinal Cord Injury.

Aileen Anderson, University of California, Irvine: ...neural stem cell lines to predict in vivo efficacy for chronic cervical spinal cord injury.

Hans Keirstead, University of California, Irvine: hESC-Derived Motor Neurons For the Treatment of Cervical Spinal Cord Injury.

Martin Marsala, University of California, San Diego: ...paraplegia: modulation by human embryonic stem cell implant.

Martin Marsala, University of California, San Diego: Induction of immune tolerance after spinal grafting of human ES-derived neural precursors.

Jane Lebkowski, Asterias Biotherapeutics: Evaluation of Safety and Preliminary Efficacy of Escalating Doses of GRNOPC1 in Subacute Spinal Cord Injury.

Arnold Kriegstein, University of California, San Francisco: Human ES cell-derived MGE inhibitory interneuron transplantation for spinal cord injury.

Mark Tuszynski, University of California, San Diego: Functional Neural Relay Formation by Human Neural Stem Cell Grafting in Spinal Cord Injury.

Jane Lebkowski, Asterias Biotherapeutics: Phase I/IIa Dose Escalation Safety Study of AST-OPC1 in Patients with Cervical Complete Spinal Cord Injury.

Bennett Novitch, University of California, Los Angeles: Molecular Characterization of hESC and hIPSC-Derived Spinal Motor Neurons.

Sarah Heilshorn, Stanford University: Injectable Hydrogels for the Delivery, Maturation, and Engraftment of Clinically Relevant Numbers of Human

Induced Pluripotent Stem Cell-Derived Neural Progenitors to the Central Nervous System.

Aileen Anderson, University of California, Irvine: Role of the microenvironment in human iPS and NSC fate and tumorigenesis.

Leif Havton, University of California, Irvine: Repair of Conus Medullaris/Cauda Equina Injury using Human ES Cell-Derived Motor Neurons.

Leif Havton, University of California, Los Angeles: Development of a Relevant Pre-Clinical Animal Model as a Tool to Evaluate Human Stem Cell-Derived Replacement Therapies for Motor Neuron Injuries and Degenerative Diseases.

Binhai Zheng, University of California, San Diego: Genetic manipulation of human embryonic stem cells and its application in studying CNS development and repair.

David Schaffer, University of California, Berkeley: Scalable, Defined Production of Oligodendrocyte Precursor Cells to Treat Neural Disease and Injury.

Click on this, for some up-to-date info on the Asterias Biotherapeutics effort.[4]

These are of course quite useful for researchers, who will find a ton of value there.

And for non-scientists, like myself? In addition to the blog entries (the understandable ones!) there is also summary info like the following.

"…On November 3, 2016, Asterias successfully dosed the first AIS-A patient with 20 million cells of AST-OPC1. No serious adverse events… have been observed in any treated patient to date…subjects in the study have demonstrated improvements in their ability to independently perform activities of daily living including feeding themselves, drinking, texting, sending emails, and signing their names…"

www.cirm.ca.gov is a library of advanced and basic information, pieces of the puzzle of cure, all available to you — for the trouble of clicking a button.[5]

---

[4] https://www.cirm.ca.gov/our-progress/awards/phase-iiia-dose-escalation-safety-study-ast-opc1-patients-cervical-sensorimotor

[5] https://www.cirm.ca.gov

17 Speaking Before Those Who Oppose

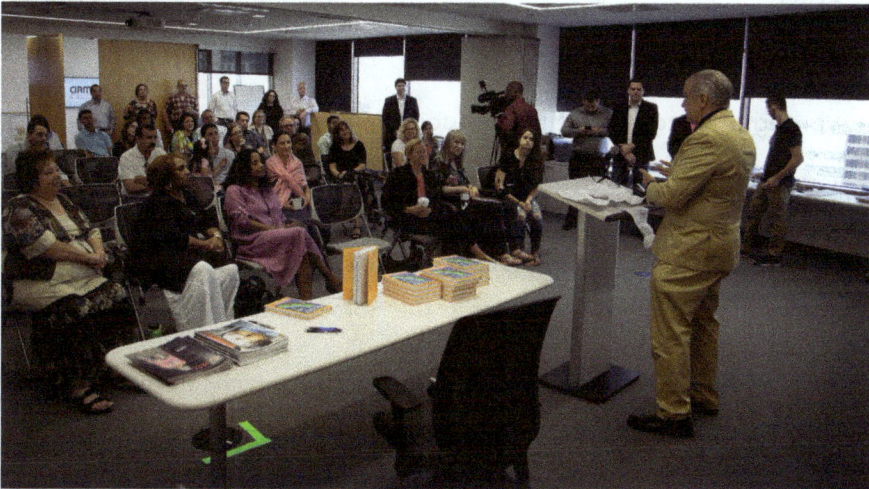

Author chats with friendly audience in CIRM headquarters — but not every audience will be on your side…. (CIRM photo.)

Those who oppose embryonic stem cell research often do so from a religious perspective. It is not an either–or situation, though, because some faiths support it (Judaic), some are against (Southern Baptist), and some take a mixed position (American Baptist).[1]

I am not a member of any organized religion, but my family is. My Dad is the son of a Minister, Gloria is emphatically Catholic. I go with her to Church most Sundays, and the roof does not cave in.

---

[1] https://www.pewforum.org/2008/07/17/religious-groups-official-positions-on-stem-cell-research/

But when she asked me would I talk about stem cells before her Bible study group, I was a little nervous.

During the 2003 campaign for Prop 71, I had an unfortunate experience with an opponent of stem cell research…A fiery-eyed woman had approached me and my sign, which read: GET STEM CELLS ON THE BALLOT! She listened to me talking to a prospective supporter, and when I turned to her, she literally <u>spat</u> on my shoe and walked away.

However, a major poll showed 72% of Catholics support embryonic stem cell research[2] although the official position of the Church itself is opposed.

Plus I admire Pope Frances, who seems a kind and caring man.

There were about thirty that night, gathered in a half-circle of folding chairs.

I brought along a favorite photograph, a picture of a "precursor", a cell in the process of becoming a motor nerve, blown up about a gazillion times. Bob Klein gave me that picture, and I generally keep it hanging on my wall.

I told them the story of our son Roman Reed, paralyzed in a college football game, and how he inspired the Roman Reed Spinal Cord Injury Research Act of 1999.

I was about to tell them about Proposition 71, which led to (CIRM) the California stem cell program.

Just then, two hands in the back of the room flew up. Their words, tone of voice and body posture seemed an invitation to battle.

Did I support adult or embryonic stem cell research? Uh-oh, here we go.

I took about ten minutes to answer that one, explaining that CIRM used a variety of stem cell procedures, also making clear (hopefully) where embryonic stem cells came from. (I avoided the mistake of saying "this is not killing babies", because as soon as you say that, the only part the audience will remember are the last two words.)

When a couple uses the In Vitro Fertilization (IVF) procedure, there are generally 15–20 leftover blastocysts remaining, which would otherwise be flushed, incinerated, or put in the landfill. Instead, the couple might donate some of the blastocysts to research. As long as there was no implantation in the womb, it was biologically impossible to make a child.

---

[2] http://www.catholicsforchoice.org/catholics-for-a-free-choice-supports-proposition-71-the-california-stem-cell-research-cures-initiative/

I went back to my lecture — but the second hand went up.

Had CIRM done any *FDA-approved* clinical trials? The words were harsh, as if we were a bunch of snake oil salesmen.

Yes, about fifty.

Had CIRM provided any *honorable* stem cell treatments?

That word "honorable" ruffled my feathers.

Now I believe in courtesy. And the group had gone out on a limb inviting me.

I was nice, of course.

I told them about the Vatican's conference on stem cell research, Unite to Cure, and how CIRM President Randy Mills had been invited (by the Pope) and was featured speaker.[3]

And as for the combination approach of stem cells made from a child's bone marrow, genes adjusted and then put back — that was literally life-saving, as in the "Bubble Baby" therapy that had saved the lives of fifty children with the immune system problem. "Remove, revise, replace", was that procedure to me.

I gave them every positive I could think about non-embryonic stem cell research.

But the two in the back kept hammering at me, one negative after another.

Finally I said:

"I have tried to be as fair and positive about adult stem cell research as I can. Where it is good, we should use it, and we do."

"But I will not tell you I am wrong, because I'm not."

"I knew this might be a hostile audience — here I was interrupted with: "No, no!" — But I came because roughly one in two Americans has a chronic disease, meaning they will never get better. One of those people is my son, who is paralyzed. One of the most promising possibilities of easing paralysis involves embryonic stem cells. It won't help Roman, because it is only for new injuries, but it has already restored hand and arm motion to several profoundly paralyzed individuals. Possibilities like that should be explored."

Afterwards, virtually the entire audience came up. They formed a line, and I braced myself — but they had genuine questions, not attacks. Almost all of them had a variation on the same question: "My (family member) has (a chronic disease) — is anything being done to fight that

---

[3] https://vaticanconference2018.com/c-randal-mills-phd/

condition?" I was delighted to refer them to the CIRM website (www. cirm.ca.gov) as well as to share what little knowledge I had.

So was it worth it?

Of course.

But in the future I will try to only take questions after the prepared portion of the speech. That way I can get my message across before the opposition begins.

18 The First 500 Pounds — And Donald Kohn

Is there a medical equivalent to the first 500 pound overhead lift? (Pinterest photo.)

Does every endeavor have a special milestone, a barrier for only the greatest champions: like the first sub-four-minute mile in track, by Roger Bannister?

In weightlifting, it was the first 500 pounds, the fabulous quarter of a ton. Who would be first to make this historic lift, to put 500 overhead?

I was a second-class weightlifter (best overhead lift: 345) in York, Pennsylvania, Spring, 1968, when Bob Bednarski woke me with a phone call.

"Come to the gym — nobody wants to train with me!", he said.

Which was understandable, it being midnight.

But this was Bob Bednarski, perhaps the greatest weightlifter who ever lived. At 250 pounds, he was outlifting men one hundred pounds heavier. Plus, he was just recovering from a broken arm![1]

Of course I said yes, and drove my ancient 1952 Dodge to the York Barbell Club, America's most famous weightlifting gym, to which "Barski" had the key.

He was working on the overhead portion of the clean and jerk, taking the bar from portable squat racks and ramming it nine feet in the air.

Two hundred twenty-five pounds was a warm up for him and he did a few quick repetitions before adding another ninety pounds — 315 for a triple, then a single with 405, four heavy plates on each side.

455 looked easier than 225 — and 475? Bar across the front of his shoulders, he dipped at the knees, shoved hard, then flung himself beneath it — swiftly separating his feet, left knee bent in front, right leg straight behind him — the weight plates made a solid reverberating clank...technical perfection.

What next?

"I could do 490, that would be a record, more than Zhabotinski ever did?"

"No, no, you are magic tonight, you can do anything," I said, "Take 500 — the quarter of a ton, first time in the history of the world!"

So easy to recommend 500 pounds for someone else to lift...

And then, he was not listening anymore. He had gone to that place where only champions can go.

Picture it. Individual squat racks on the platform, bar across them. What a collection of weights. On each side were 4 big 45 pound plates (405), then a 25 (455), a 10 (475), a nickel or 5 pound plate (485), a 2 and a half (490), and the 5 pound iron collars, not too tight — Bednarski liked them a little bit free — 500.

He chalked his hands, then paced back and forth behind the low wooden platform with its heavy reinforced rubber mats: once, twice, he paced, three times, four; and just when I thought he was never going to try it — he stepped onto the platform. He approached the squat racks,

---

[1] http://www.chidlovski.net/liftup/l_athleteResult.asp?a_id=345

took the bar across the front of his shoulders, between deltoids and neck. He shuffled forward, the bar so heavy, it visibly bent.

He breathed in and out, rapidly, concentrating, focusing every atom of his being.

He dipped at the knees, slower than you might think at first — then invisibly quick, hyperdrive, ramming the bar overhead, arms straight, feet front and back.

Could I believe what my eyes told me?

The bar was overhead: a quarter of a ton. Bednarski recovered his feet perfectly, shuffling the front foot six inches back, then back foot joining front, side by side.

He was roaring as he stood there, 500 pounds overhead. It was his, he could hold it as long as he wished.

And then, he let the weight drop, no attempt to control its fall. I heard the *whoosh* of bar through air — and then the crash.

The bar was alive. One side hit the reinforced rubber mat, landing so hard the bar rebounded, crashing, then leaping off the platform.

When it hit the concrete floor, the impact gouged a chip, and as the bar bounded across the floor, it dug progressively shallower chips, like footprints to mark an historical event...

In regenerative medicine — stem cell research and gene therapy — is there a similar clear-cut milestone: a stem cell goal to excite the mind and stir the senses, of which one might say, "Spectacular!"

Shinya Yamanaka earned the Nobel Prize for inventing the technique called IPS — Induced Pluripotent stem cell research — (iPSc) an imitation of embryonic stem cell research.[2]

And of course John Gurdon did amazing work on Somatic Cell Nuclear Transfer (SCNT), a form of cloning for cells.

But I have a technique in mind, which might be called the "three R" technique: **Remove, Revise, Replace**. That is just my layman's term for it, of course. I am no scientist.

But that technique has cured patients with two kinds of "incurable" diseases, and may provide healing for more.

As with all medical advances, many scientists have contributed to its development: but one person stands out:

---

[2] https://www.eurostemcell.org/nobel-prize-goes-stem-cell-pioneers-john-b-gurdon-and-shinya-yamanaka

Donald Kohn, of the University of California at Los Angeles. He is small-built, like a gymnast, full of energy and strength. He works at the Eli and Edythe Broad Institute, as well as at a corporation called Orchard, Inc.

His science has quite literally saved lives.

"Over the course of 30 years of research, Kohn has developed new clinical methods to treat genetic blood diseases using blood stem cells that have been modified to remove genetic mutations."[3]

It works like this: he <u>removes</u> blood stem cells from the bone marrow of patients with a disease, then he <u>revises</u> the genes that are bad, and finally <u>replaces</u> them into the patient. **R**emove, **R**evise, **R**eplace — the Three R technique.

Want some examples?

#1: I am sure you remember Evangelina Vaccaro, the little girl cured of SCID (Severe Combined Immune Deficiency) better known as the "Bubble Baby" Syndrome. Dr. Kohn's technique cured Evie — and 49 other children who would otherwise almost certainly have died of that condition.

#2: Maybe you know Brenden Whitaker, a young man cured of the normally fatal granulomatous disease? He is alive, thanks to Dr. Kohn and the Three R technique.

#3: If you read one of my previous books ("CALIFORNIA CURES"), you might remember a vicious condition called IPEX, a combination of diseases (Immune Dysregulation, Polyendocrinopathy, Enteropathy, X-linked), generally fatal.

Here I must give a shout-out to Dr. Rosa Baccheta of Stanford, who has been working more than 20 years to defeat IPEX. When the cure comes, her fingerprints will be all over it.

But back to Donald Kohn and the "Triple R" technique...

Dr. Kohn, Katelyn Masiuk and other UCLA researchers are testing (in mice) a way to "reverse the genetic mutation that causes a life-threatening autoimmune syndrome called IPEX...The gene therapy is similar to the technique Kohn has used to cure patients with...SCID."[4]

---

[3] https://stemcell.ucla.edu/member/kohn

[4] https://www.sciencedaily.com/releases/2019/01/190110171056.htm

Donald Kohn's stem cell/gene research "triple-play" should go down in history.... (UCLA photo.)

#4: He may even be able to use the technique against sickle cell anemia.

"Kohn is currently conducting clinical trials that test similar stem cell gene therapy techniques for other diseases, including sickle cell disease ... the most common inherited blood disorder in the U.S."[5]

What a struggle it has been. Here is a note from Dr. Don:

"We have been working on developing stem cell gene therapy, with CIRM support, for almost 10 years."

"Under a Disease Team I grant, we performed the necessary ... work on a ... vector that expresses a modified version of ... beta-globin that has anti-sickling properties (emphasis added — DR) to apply to FDA for an IND for the Phase I trial ..."

"We got the IND (Investigational New Drug FDA permission to go forward) in 2014 and treated one patient in 2015. Unfortunately, the level of gene insertion into stem cells was too low to be beneficial. We spent the next several years modifying the vector ... and identified some 'Transduction Enhancers', adding to the culture, increasing by 10-fold the level of gene transfer."

---

[5] http://newsroom.ucla.edu/releases/family-travels-to-ucla-save-son-with-bubble-baby-disease

"We submitted an IND (Investigational New Drug) amendment to FDA for these changes and were approved about two months ago."

"A first sickle cell patient was consented into the modified study last month and is undergoing eligibility testing. We have at least 20 or more other sickle cell patients who have been referred to our trial either by their doctors or by themselves. The trial calls for a total of 6 patients. We hope to treat 1–2 this year and the remainder next year."

"In the meantime, we are also working on using CRISPR to correct the sickle mutation in stem cells with Mark Walters and colleagues at UCSF and UCB. This just received a CLIN1 award (from CIRM) to do the IND enabling studies and we hope to be in the clinic within 2 years."

— Donald B. Kohn, M.D., personal communication.

And who got the first 500? At the 1968 Senior National Weightlifting Championships that year, Bob Bednarski set two World Records: a 456 ½ pound Olympic press, and a 486 ½ clean and jerk. Unofficially, he later cleaned 490 (floor to shoulders) and jerked an incredible 525 from the racks. But he never got the first 500.

That honor went to Vasily Alekseyev in 1970. I was there that day, at the World Championships in Columbus, Ohio, when the huge man hoisted the first 500. He did it smoothly, very business-like. When he set the bar down, and the crowd went insane, Alekseyev did not seem to understand the fuss. He smiled when a beautiful girl (a Russian gymnast) ran onto the platform and leaped into his arms, that part he got. But the 500 pound lift? It was just the 17th World Record he had done that year (he eventually did 80) — and he thought in kilos (227.5), not in pounds...[6]

---

[6] https://en.wikipedia.org/wiki/Vasily_Alekseyev

19 Why Fetal Cell Research Must be Allowed

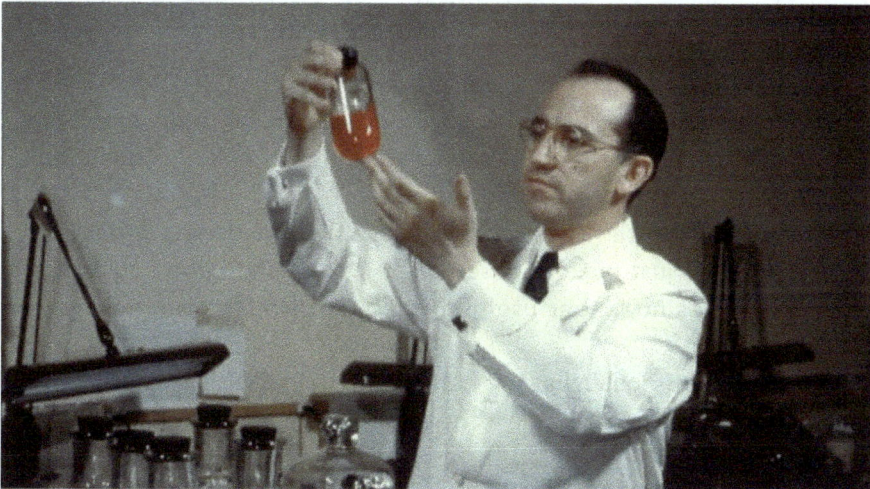

When Jonas Salk invented the polio vaccine, which has saved countless lives, he was using fetal tissue.... (Photo by History.com.)

The Trump administration recently issued an order blocking funds for fetal cell research, and may be planning to ban the procedure altogether.[1]

Why would they want to eliminate this form of medical research?

Fetal cell research involves tissues from an abortion.

No one, I think, *wants* the tragedy of an abortion.[2]

---

[1] https://www.sciencemag.org/news/2019/06/trump-administration-restricts-fetal-tissue-research

[2] https://en.wikipedia.org/wiki/Abortion

But it happens, nonetheless. Around the world, an estimated fifty-six million abortions take place, every year.[3]

A grim question: imagine you are a surgeon, in the operating room. Before you are two babies — one dead, the other dying — if a tissue sample from the dead would save the life of the living, would you hesitate?

That is the question on fetal cell research. The cells used come from an abortion, which all regret. But if the abortion has already happened, is it wrong for something life-saving to come from it?

Think of organ donation.

For scientists/doctors to develop cures, they need cadavers — donated bodies and organs — both to learn from, and to use for transplants.[4]

When my cousin died in a car crash, his mother had to make the wrenching decision to donate his organs. But years later, she met a man, alive today because of the gift of her son's heart. After his own death, my cousin saved somebody's life.

After I die, I want any part of me still useful to be donated to science. Organ donation just makes sense. What if someone blind could regain vision from eyes no longer useful to me? Or gain life from my liver, or heart?

Back to fetal tissue. What if it had always been illegal?

Jonas Salk's polio vaccine was developed, tested and grown on fetal tissues.

Done since the 1930s, fetal tissue research has not only wiped out polio, but also eliminated rubella — which once killed 50,000 babies a year through miscarriages. Fetal tissue was used for vaccines against measles, mumps, chickenpox, whooping cough, tetanus, hepatitis A and rabies — all conditions which once killed people.

Today, fetal cell research is being used to fight the Zika virus, as well as to attempt therapies which may eliminate HIV/AIDS. Fetal tissue research may also help end Early Pregnancy Loss — miscarriages which cost the lives of nearly a million infants every year.[5]

"Since 1994, these vaccines saved society an estimated $1.38 trillion dollars."[6]

---

[3] https://www.guttmacher.org/fact-sheet/induced-abortion-worldwide
[4] https://register.donatelifecalifornia.org/register/
[5] http://www.hopexchange.com/Statistics.htm
[6] https://selectpaneldems-energycommerce.house.gov/our-work/benefits-fetal-tissue-research

Massive regulation controls the obtaining and use of fetal tissue. There can be no profit in its sale (beyond reimbursement of costs), and it can only be authorized once an abortion has been selected. A very clear explanation of the various rules was put together by the Congressional Quarterly.[7]

But *should* we do it? Wisconsin Bioethicist Alta Charo put it best:

"We have a duty to use fetal tissue for research and therapy...Virtually every person in this country has benefited from research using fetal tissue...every child who's been spared the risks and misery can thank the...scientists who used such tissue in research yielding the vaccines that protect us...fetal tissue research (has) saved the lives and health of millions of people."[8]

Here is a September, 2018 letter of support for the research, written by the American Association of Medical Colleges, with multiple group support.[9]

Finally, here are some very short selections of support, drawn from groups. The titles are mine, the contents are drawn from their long individual letters.[10]

Why We Need Fetal Tissue Research?

"Joined by more than 60 scientific and academic organizations, the AAMC stands by its statement, that: restrictions on fetal tissue research...would limit new research on vaccines not yet developed, for treatments not yet discovered, for causes of diseases not yet understood." — Darrel G. Kirch, MD, President and Chief Executive Officer, American Association of Medical Colleges.

Fetal Tissue Research (FTR) Saves Lives:

"The use of fetal tissue has been an important factor in the development of numerous lifesaving vaccines for children (including)...chicken pox, hepatitis A, polio, rabies, and rubella...it is estimated that for just children born between 1994 and 2013, vaccination will prevent an estimated

---

[7] https://fas.org/sgp/crs/misc/R44129.pdf
[8] http://www.nejm.org/doi/full/10.1056/NEJMp1510279#t=article
[9] https://www.aamc.org/download/492010/data/communitylettertocongressionalleadershipregardingfetaltissueres.pdf
[10] https://cdn.cnsnews.com/attachments/exhibits_from_chapter_9-select_investigative_panel-final_report.pdf

322 million illnesses, 21 million hospitalizations and 732,000 deaths over the course of their lifetimes..." — Bernard P. Dreyer, MD, President, the American Academy of Pediatrics (AAP), a non-profit professional organization of 64,000 primary care pediatricians.

Preserves Women's Health:

"Representing more than 57,000 physicians and partners in women's health, the American Congress of Obstetricians and Gynecologists (ACOG) is committed to fostering improvements in all aspects of the health care of women...Although ACOG does not conduct fetal tissue research, we recognize the value of medical research in improving the lives of our patients. As an organization of physicians and health care practitioners who work every day to save and improve women's lives, (we) strongly support...such research..." — Mark S. DeFrancesco, MD, President, American Congress of Obstetricians and Gynecologists.

Treats Diseases for Young and Old:

"The use of fetal tissue for medical research has...allowed scientists and physicians to develop new treatments for both adult and pediatric illnesses, because disease-causing mutations target fetal cells specifically." — Peter M. Grollman, Senior Vice President, The Children's Hospital of Philadelphia.

May Help Protect the Developing Brain:

"Research using fetal tissue makes it possible to examine how viruses and other ailments affect a developing brain. Using (fetal tissues) scientists can study brain tissue in the petri dish and better understand how viruses and other diseases progress and how best to stop them..." — Dr. Hans Snoeck, Professor of Medicine in Microbiology and Immunology at Columbia University Medical Center.

Johns Hopkins Research on Als:

"As nerve cells degenerate (from ALS) the muscles they control...stop working and ALS patients typically die of suffocation...Injecting a certain type of fetal cell into mouse ALS models appears to protect the existing cells from degenerating..." — Paul B. Rothman, M.D., CEO, Dean of Medical Faculty, Johns Hopkins Medicine.

Applicable to Many Conditions:

"Fetal cell lines have been used in medical advances...including an arthritis drug and therapeutic proteins that fight cystic fibrosis and hemophilia. Understanding degenerative diseases such as Alzheimer's, Huntington's, and a host of other devastating...conditions, depend specifically on access to fetal tissue. Ongoing fetal tissue research is critical for continuing advances in regenerative medicine, including organ/tissue regeneration of heart, liver, pancreas, lung, muscle, skin, and more, holding out hope for a wide variety of therapeutic discoveries." — Harriet S. Rabb, Vice President and General Counsel, Rockefeller University.

Helps Develop Animal Models for Numerous Conditions:

"This type of research has helped improve our understanding of numerous health issues including early brain development, neurocognitive disorders, maternal and fetal health conditions, congenital heart defects, Down syndrome, and other infectious diseases..." — Daniel Dorsa, PhD, Senior Vice President for Research, Oregon Health and Science University.

Positive Findings Have Led to Clinical Trials:

"... Clinical trials with cells from fetal tissues are ongoing for age-related macular degeneration (blindness) and chronic liver disease." — Michael D. Amiridis, Chancellor, University of Illinois at Chicago.

Impacts of Environmental Toxins:

"One researcher...procured fetal tissue...to conduct research (on) the impact of environmental toxins, to (guard)people from environmentally-induced disease..." — Cynthia Wilbanks, Vice President, Government Relations, University of Michigan.

<u>Reduces</u> Need for Abortions:

"Research using human fetal tissue...was critical...to prevent mother-to-child transmission of HIV. That research has saved over 1 million infants in the past 10 years, while reducing elective abortion in HIV-positive women by more than half." — Brooks Jackson, MD, MBA, Dean of Medical School, University of Minnesota

May Lessen Abnormalities:

"One of the most promising areas of research involving the use of human fetal tissue is the study of how and why abnormalities develop in a fetus. Congenital abnormalities occur in 3–4% of pregnancies and cost society billions of dollars per year in long term disability care, and incalculable emotional toll on patients and families. Human fetal tissue research enables scientists to learn...how abnormalities arise...(and may) lead to treatments to decrease the frequency of the anomalies and...may improve the quality of life for patients." — Susan E. Phillips, Senior Vice President for Public Affairs, University of Pennsylvania Health.

Develops Knowledge Against Cancer:

"(Fetal) cell line WI-38...(resulted in) 1,387 published studies focusing on...cancer, cell growth, cell aging, anti-cancer drugs..." — Robert N. Golden, MD, Dean, School of Medicine, University of Wisconsin-Madison.

Fetal Research Previously Evaluated:

"In 1988, the Human Fetal Tissue Transplantation Research Panel...provided an important public forum to consider the scientific value and ethical acceptability of fetal tissue research, hearing testimony from lawyers, ethicists, religious leaders, biomedical researchers, clinical physicians and the general public, including families with children afflicted by disease and disability. ...the panel determined that the research was in the public interest with the potential to help millions...To abandon this approach would significantly impede...the alleviation of human suffering..." — Robert J. Alpern, MD, Dean of Medicine, Yale School of Medicine.

To whom is our greater obligation: the already deceased? Or those whose lives may be saved?

20 "Told Your Child is Going to Die..."

This tiny being, embodying our hopes and dreams of cure. (CIRM photo.)

In the 25 years since my son became paralyzed, I have met a lot of great people: dedicated warriors for research, advocates for patient rights, caregivers whose involvement with a loved one often keeps him of her alive: people who fight years and years hoping that cure will come. None of us knows for sure, there are no guarantees, yet we dare not give up — because maybe if we hang on just a little bit longer...

But what would it be like, if we won — and got the cure?

Here is a brief interview with somebody who did just that: Alyssia Vaccaro, mother of Evie, a child born with "Bubble Baby" syndrome.

DR: When did you first know, that something was wrong?

AV: As a mom I had a gut feeling that something was wrong, almost immediately. Both girls were born early, but Evie spent a lot of time in the Neonatal Intensive Care Unit — NECU. Her skin was super red; we used to call her Lady Bug, she was so red. Our other twin, Annabella, had normal coloring. Also, Evie would spit up more. Her older sister could be on room air; Evie could not. On day 10 we got them both home, but continued to worry. Sometimes Evie would turn gray, and she always seemed uncomfortable. She had acid reflux. We were endlessly comparing, questioning, why was one twin doing better than the other? The pediatrician kept constantly rechecking her blood's white cell levels — cells for fighting infection.

When Evie was 20 days old, they said her white blood cell levels were too low, and we had to readmit her into the hospital. But we had to stay in the car outside until they had a bed ready for her — We could not just carry her into the emergency area, because of germs.

We did not know she had Severe Combined Immune Disorder (SCID). We tried to rule out other conditions like blood cancer, and some tests were inconclusive and some brought false results. We checked out of the hospital (again) when she was 25 days old. The day after that, we received a call requesting more blood — which tested positive for SCID — And at 35 days, we took her back to the hospital again.

DR: How did you feel?

AV: How could any mother feel — when you are told your child is going to die? We learned about the "bubble baby." My husband Christian and I had seen that ridiculous John Travolta movie, and got to know David Vitter and the real-life people behind the story. We felt guilty that we had laughed at the Seinfeld show, where they joked about a boy in a plastic bubble. We needed bone marrow. It was the worst feeling in our life, as if the ground was pulled out from under our feet.

DR: How does it feel now, after your daughter became the first person in the world to be actually cured?

AV: Just telling the story re-introduces the fear. To say the words "cured child" triggers actual nausea. I am always waiting for the other shoe to drop. When the doctors ask, is she up to date on vaccines, I can say yes, and I am so glad I can send her to school. Her immune system can fight a cold now. But the fear remains.

DR: Do you stay in touch with Dr. Kohn?

AV: Oh yes, for one thing I am glad to be a patient advocate for him and Orchard Therapeutics. We are friends now, and I do not only send him questions about our daughter's symptoms. We no longer just say, "We are ready to do the measles vaccine, what should we look for?" We also exchange Christmas cards.

DR: What does Evie like to do?

AV: She loves to ride horses, and play tennis; she has a great backhand! She also loves the beach and just being outside; digging up worms after a storm. She and her twin get along maybe 70% of the time — 5% they actively dislike each other — and 25% they are indifferent! Evie is also very considerate of her sisters — whenever she gets something, she will say, "what about my sisters?".

DR: When I first met Evie, I was actually a little afraid of her, because here was this tiny little being in her Supergirl costume, and she embodied the life and death struggle and the hope for cure for so many people. How do you explain to her, how important she is, and why?

AV: No idea! We have not really explained it to her yet — when it is time, and she wants to ask questions, we might show her that stupid movie. But what will probably happen, around the age of 8, she will just Google her name, and find out.

In the meantime, when she asks why so much blood has to be taken from her arms, we tell her the truth, which is that she is helping cure others, so they will have a chance to survive. Technically she is in a long-term clinical trial — scientists study her, in order to cure other children. But Evie just says she has unicorn blood!

DR: How is her health, otherwise?

She has hearing difficulties. At first, we thought she might need speech therapy, and enrolled her in classes. She does have trouble with "S" sounds, for instance, like she might say "nake" instead of "snake." But it turns out a lot of people with SCID have trouble with hearing as well. So she has hearing aids, and is learning that words are different from what she first thought they were.

DR: It does not seem fair, to have both SCID and hearing problems as well!

AV: There are worse things in life — and she still rocks her world.

DR: She does indeed! Question: what are your thoughts on CIRM?

AV: My husband Christian and I agree, we can't be given something, which we were, and not try to give back, to help others. How could we not try to get more funding for CIRM? No one is safe from chronic disease. It will happen again, maybe to someone I love, or it might be me. Regenerative medicine is in its infancy right now, but we are already seeing it work.

Right now, it is expensive. But over time, cure therapies will become cost effective, cheaper for everyone.

Think of a flat screen TV. They used to cost around $7,000, but now you can get one at CostCo for $200. Over time, you perfect it, and costs will be reduced.

CIRM's success is just beginning.

This is the sunrise.

21 Flat Feet and Neuropathy

Robert Baloh: the man who may reduce foot pain! (UCLA photo.)

Other than being spectacularly flat, my feet gave me no trouble for their first 72 years. True, when wet, the tracks of my feet would be completely filled in — no arch. And the noise of my footsteps was "flop flop flop!", like walking in swimfins. I was even told, when I took my Army physical, that the flatness of my feet would disqualify me from the draft! But I was a volunteer, signed up to serve, and went anyway.

Army doctors actually tried to repair my feet, wrapping them in hot plaster casts, making molds for orthotics, plastic inserts for my shoes. At one point, I was given six weeks' restricted duty, a "buck slip" prohibiting me from walking more than 100 yards. About all I could do was walk to the gym and lift weights, which I loved, or saunter to the mess hall and eat the Army's alleged food.

Presently the doctors got bored and said I was fine, which I could have told them already. My feet were still flat, but the only time they hurt was when I put those plastic things into my boots.

So I threw the inserts away, and quit thinking about my feet.

Until about a year ago...

In the Spanish Inquisition, there was a torture called the strappado, where the soles of the feet were beaten with sticks. Apparently nerves in the body end up in the feet, and to be struck there was said to be excruciating.

Why do such thoughts occur to me now?

For no discernible reason, my feet developed numbness and pinpricks; at first just odd sensations, like wearing super-tight socks. Then it got worse.

"Peripheral neuropathy," said the doctor, "PN".

It is not rare. Twenty million Americans have PN right now, according to the National Institute of Neurological Disorders and Stroke (NINDS).[1]

There are over 100 varieties of this nerve-damage disease. It is not life-threatening, but some varieties (Guillain–Barré Syndrome) bring paralysis. Some are side effects of larger diseases, like diabetes: some are caused by trauma.

My son Roman has it in his fingers, as a result of his spinal cord injury. Some are nerve-insulation diseases, where the myelin nerve-covering is not right. Some kinds go away over a period of years. Others stay with you forever.

My son's advice? "Get used to it, Dad." I did not find that comforting, somehow.

Pills (non-narcotic ones like gabapentin) sometimes moderated the symptoms. Other times it felt like eagle claws clutching in.

And it does not go away. Dealing with the not-quite-pain is a constant struggle, like holding my breath. If I relax my guard, I might snap at Gloria, which is not advisable. Also, sometimes in my sleep, she says, I make moaning noises until she pokes me in the shoulder sufficiently.

Lately, the condition is expanding territory. It started with my feet, then the ankles, calves, and now it seemed to want to be everywhere.

---

[1] https://www.ninds.nih.gov/Disorders/Patient-Caregiver-Education/Fact-Sheets/Peripheral-Neuropathy-Fact-Sheet

"Oh yes," said the doctor cheerfully, "it can spread up your legs, and then jump across to your arms."

And sometimes I feel faint tinglings beginning in my face.

Can it be fixed?

Some good people are trying to find out.

Like Dr. Robert Baloh, of the University of California at Los Angeles.[2]

Working with a grant ($3,031,737) from CIRM, Dr. Baloh is attempting to "develop (stem cell) lines from patients with...peripheral neuropathy..." This will give him a "disease in a dish" that he can work with.

After that he hopes to: "genetically correct the defect in (the) cells... and (turn) them into Schwann cells, which will (insulate) the (nerves)..."

How is he doing? Quoting from his CIRM document:[2]

"Overall we have made excellent progress...We have successfully generated (stem cell) lines from...patients and characterized them (turning them into the needed Schwann cells)."

"We found that (the cells) could engraft to form myelin...in (rat models similar to) human neuropathies."

"This demonstrates for the first time proof of concept for...(a) cell therapy for diseases of peripheral nerves outside of simple nerve injury."

"Remaining hurdles include the lack of...technology for generating human...(stem cell) precursors for transplantation that (can) generate myelin..."

As I understand it, the problem is with the insulation (myelin) around the nerves. But now the scientists hopefully will be able to make stem cells from a human with PN, put the condition inside a lab rat, and then figure out how to repair it.

Selfishly, I would like scientists to develop what is needed to alleviate my personal peripheral neuropathy.

Or, to put it into scientific terminology:

"Fix my feet, Doc!"

---

[2] https://www.cirm.ca.gov/our-progress/awards/human-ipsc-modeling-and-therapeutics-degenerative-peripheral-nerve-disease

22 Battling Schizophrenia

"In a lot of psychiatric diseases, there is dysfunction in the connection *between the cells*," – Rusty Gage. (Salk Institute photo.)

Imagine a calm persuasive voice, telling you to pick up a gun, put it to your head, and *squeeze the trigger* — you would say no, of course — but what if that voice repeated itself over and over, and it came from inside your mind?

Schizophrenia "afflicts an estimated 3 million Americans (with) hallucinatory symptoms, such as hearing voices. Lifelong drug treatment...can help keep symptoms in check, but there is no cure — and ten per cent surrender to the disease by taking their own lives."[1]

---

[1] https://blog.cirm.ca.gov/tag/schizophrenia/

What might it be like to live with schizophrenia? Check out this three minute video.[2]

What a difficult and complicated problem!

But the challenge must be met. Several top scientists and their labs are struggling to understand and fight schizophrenia. Here is my non-scientist's interpretation of three pieces of the puzzle: work from Marius Wernig, Christina Chatzi, and Fred Gage, who are using different stem cell-based approaches to target the disease.

First, what causes schizophrenia? Could it be a gene that activates parts of our body during development? Or maybe the opposite, deactivating rival genes — so one survivor can be the strongest?

In nature, the cuckoo bird approaches a robin's next, and puts its own egg among the others. When the cuckoo chick hatches, it is much bigger than the others, and eats more than its share — and the other chicks may actually starve.

Like the aggressive cuckoo, some genes responsible for nerve development become activated — while other genes remain silent.

Marius Wernig, MD, associate professor of pathology at the Stanford University School of Medicine, points to a protein called Myt1l (mighty-one L), which "block(s) the activation of genes related to lung, cartilage, heart and other(s). The one set of genes that Myt1L repressor did not act on was neuron-specific genes." By turning off the genes for the other cell types, Myt1l pushes development of the nerve cells.

However, Wernig says, if there are problems (mutations) with Myt1L "...neural (nerve) cells get a little confused."[3]

Such "confusion" may be devastating.

"Myt1L mutations have been recently found in people with autism, schizophrenia, major depression, and low I.Q..."[4]

While working at Sanford-Burnham, Christina Chatzi questioned: if one kind of neuron causes a mental illness, could a second kind of neuron inhibit the first? And if so, how do we provide the needed neurons?

---

[2] https://www.youtube.com/watch?v=IehtMYlOulk (video)
[3] https://med.stanford.edu/news/all-news/2017/04/nerve-cells-actively-repress-alternative-cell-fates.html
[4] https://blog.cirm.ca.gov/2017/04/05/cirm-funded-team-uncovers-novel-function-for-protein-linked-to-autism-and-schizophrenia/

Knowing that "some inhibitory neurons rely on...retinoic acid (a form of vitamin A)...Chatzi wondered if exposing embryonic stem cells to retinoic acid would result in these inhibitory neurons."[5]

Working with Gregg Duester in his lab, Chatzi found that mice who were not able to make the retinoic acid in their bodies had "a serious deficiency in (certain) neurons...(which has) been associated with several neurological disorders, including Huntington's disease, autism, schizophrenia, and epilepsy..."[6]

But what if there was a way to study — not the nerve cells themselves — but the *connections* between them?

At the world-famous Salk Institute, researchers are studying the communication between neurons by developing "mini-brain" models in a Petri dish.

"In a lot of psychiatric diseases, there's dysfunction in the connections between the cells," says Professor and senior author Rusty Gage. "But it's been difficult to study the connections between human neurons in the lab, until now."

Dr. Gage's group came up with a disease-in-a-dish model of schizophrenia: a way to study the condition without hurting a patient. Using the different dishes, one person's stem cell line per dish, the team compared the interactions of cells from people with schizophrenia and not.

One key difference? Neurons "in the schizophrenic group had dampened activity patterns and less signaling between the sets of neurons."[7]

Summarizing the three approaches, Mitra J. Hooshmand, PhD, (Director of Scientific Programs for Americans for Cures Foundation) said:

"Dr. Wernig is using cell reprogramming techniques to identify the genes involved in development of schizophrenia ... Dr. Chatzi is using embryonic stem cells to generate inhibitory neurons and study their role in the manifestation of (the disease) ... and Dr. Gage is using the 'disease in a dish' model to understand the role of neural networks in the development of schizophrenia." — Mitra Hooshmand, personal communication.

---

[5] https://www.cirm.ca.gov/blog/04122011/making-neurons-lose-their-inhibitions
[6] http://journals.plos.org/plosbiology/article?id=10.1371/journal.pbio.1000609
[7] https://www.eurekalert.org/pub_releases/2018-05/si-icl050218.php

Until we have *cure*, we must have *care*: to support those who suffer mental illness. Let's meet an expert in that area.

Alfred "Al" Rowlett serves on the board of directors of the California stem cell research program. He is the patient advocate for mental illness. He was appointed there because of his life-long work for such non-profits as the Turning Point Community Programs (TPCP), for which he is Chief Executive Officer.

Al helps arrange service for children and adults with psychiatric disabilities — throughout their entire life span.

Working with psychiatrists, case managers and therapists, Al works to provide life skills training, housing benefits, and other forms of psychiatric rehabilitation and crisis support — so people in need can avoid being hospitalized.

"If they cannot come to our office, we go to them, sometimes twice a day, every day, sometimes for medication support. If they desire to work, can we help them find a job — and plan how to get to work every day?"

"There are obstacles and barriers such as trying to locate affordable housing. Social stigma is pervasive, all too often based on false images promoted in movies and on TV. Verifiable data suggests the opposite; individuals diagnosed with psychiatric illnesses are often ostracized and homeless. Co-occurring symptoms associated with alcohol and drug use/abuse make for a deleterious outcome."

Rowlett points to mental illnesses such as schizophrenia as having "clear evidence of a biological component," meaning it may, in time, be overcome.

We need, in short, a funding program like the California Institute for Regenerative Medicine; we need CIRM. As Al puts it:

"CIRM is a place providing hope: where specific clinical interventions are sought — which may change the trajectory of a person's life." — Al Rowlett, personal communication.

"CIRM funding on stem cell technologies for disease modeling of neurological systems is essential in the development of new treatments for conditions including schizophrenia, bipolar, addiction, and other mood and cognitive disorders." — Marion J. Riggs, Founder of Loving Mind Institute.

23 Of Werewolves, Plague, and the Zika Virus

Alysson Muotri used a microscopic "mini-brain" to test medications against the Zika virus. (Facebook photo.)

I have a private theory about where the werewolf legend began.

The Black Plague killed perhaps 1/3 of the population of Europe. This created a huge burial problem. Survivors of the disease were often weak, and digging so many graves was beyond their ability.

So at night the wagons came.

"Bring out your dead, bring out your dead!" cried the drivers. The bodies were carted to the edge of town, and stacked up. Wood was piled on and around the piles of corpses, and fires slowly consumed them.

But in the forests nearby, wolves smelled the cooking meat, and they came. At first they were cautious, but the meat was plentiful, few humans had the means to fight them, and their fears vanished.

Imagine wolves, snarling, tearing at bodies, against a backdrop of roaring flames.

And sometimes, when the people's own hunger grew too great, they fell prey to cannibalism. And since many people were too busy to shave in those days…

Wolves and hairy-faced cannibals: not too difficult to see where legends of horror might have begun.

The Black Plague is no more, but other nightmares may take its place. Like the Zika virus.

Zika…Zeeekaahh…sounds poisonous and evil, doesn't it? But the reality is worse.

Some adults may go through a session of Zika and barely notice it: mild headaches, stomach upset, aching in the joints. Others may contract the paralyzing condition Guillain–Barré, or acute myelitis, or other disorders of the nerve system.

But the worst danger is to the unborn child.[1]

An infant infected with Zika may be born with an unnaturally small head, and a damaged brain. This can mean death before birth, or severe handicaps.

Alysson Muotri is a scientist whose work I have followed for years. He is a terrific asset to the cure research community, not only because of the quality of his science, but because he can explain himself in "people talk" — so important! He is best known for his efforts against autism, but right now let's focus on Zika.

Dr. Muotri works at the University of California at San Diego, UCSD. For a look inside his lab, go here[2]:

Not long ago, there was a major outbreak of the Zika virus in Brazil, Dr. Muotri's original home. Before then, Zika (originally found in Africa, beginning in Uganda's Zika forest) was not well known, or widespread.

But the genes mutated. Suddenly, people began to get terribly ill, and some died.

A Chinese study tested three early strains of Zika virus on lab mice: 17% of them died. But the new Zika? It killed them all.[3]

---

[1] https://www.webmd.com/a-to-z-guides/zika-virus-symptoms-prevention?page=2#4

[2] https://healthsciences.ucsd.edu/som/pediatrics/research/labs/muotri-lab/Pages/default.aspx

[3] https://www.webmd.com/a-to-z-guides/news/20170928/how-zika-virus-went-from-mild-to-devastating?src=RSS_PUBLIC#1

Horrifyingly, it is possible that Zika may cause Autism Spectrum Disorder.[4]

And like the Black Plague, Zika is spreading. The Aedes mosquito plunges its sharp proboscis into infected people, and then carries their virus to others.[5]

Recently, cases have been found in Brownsville, Texas; Miami, Florida; and Long Beach, California.[6]

To me, this feels like early stages in a plague.

In response, the Center for Disease Control has advised pregnant women to "delay" visiting Brownsville and Southern Florida. The World Health Organization suggests women in at-risk areas postpone their pregnancies. Already-pregnant women in danger zones are advised to protect themselves from mosquito attack, using repellent and bug spray. And as the disease can also be spread by sexual contact, condoms or abstinence may block one avenue of transmission.[7]

Attempts have been made to wipe out the mosquito species, by releasing sterile males. But as yet there is no cure.

Could stem cell research help in the fight against Zika?

First, we need to understand how the disease is spread throughout the body, like the swooshing of a poisoned mop.

Think of a janitor with a mop and a bucket — but the bucket contains poison. So everywhere the janitor mops, poison is spread. The "janitors" of the nervous system are called microglia. They devour waste products of the nerves, the leftovers of Neural Progenitor Cells (NPCs). When the microglia eat an NPC with Zika, the infection spreads throughout the body.

How might this be fought? Using skin cells from both infected and healthy people, Dr. Muotri has made stem cell colonies (lines) of NPCs and microglia. He verified that Zika was indeed "eaten" by microglia.

Next, he used a zika-infected "mini-brain," a tiny dot of nerve tissue called an organoid, to test some FDA-approved medications. (The advantage of using FDA-approved drugs means much less testing, costs and delay.)

---

[4] https://www.ncbi.nlm.nih.gov/pubmed/30630174
[5] https://www.cdc.gov/zika/prevention/transmission-methods.html
[6] http://www.newsweek.com/zika-virus-long-beach-microcephaly-aedes-aegypti-mosquito-627482
[7] https://www.nytimes.com/2016/06/10/health/zika-virus-pregnancy-who.html?_r=0

He found a drug (Sofosbira) originally made to fight Hepatitis C — but which might kill the Zika virus as well.

Using this new "model," Dr. Muotri hopes to get to human trials quickly.[8]

For an article by Muotri, describing his approach to battling Zika, go here.[9]

And if the disease mutates again? Worse may be in store.

Better to fight it now.

---

[8] https://www.cirm.ca.gov/our-progress/awards/treatment-zika-virus-infection-and-neuroprotection-efficacy

[9] https://www.ncbi.nlm.nih.gov/pubmed/29048558

24 Getting <u>All</u> the Cancer

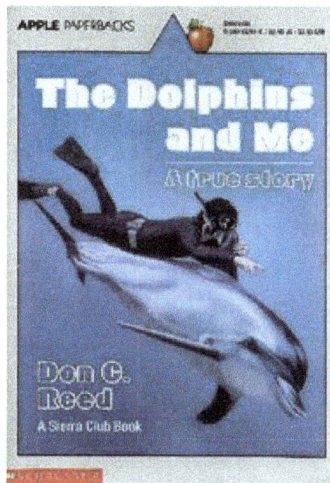

THE DOLPHINS AND ME: As a diver for Marine World, I sometimes held a dolphin still — for veterinary attentions. (Author's book cover.)

"Which arm?" asked the blood-taking nurse.

"I hear the left arm circulation is closer to the heart, and hurts less," I said.

She wrapped a plastic band around my left bicep so the vein would stand out, reached for the needle.

"Wait a minute," I said quickly, "I have a story, and a question."

She looked at me, like okay, we aren't too busy...

"I used to work as a professional diver at Marine World," I said, "and sometimes I would hold a dolphin's tail still for the vet to take blood, like you are about to do.

"But a dolphin's skin is smooth, and you cannot see the veins. So my question is: how do you know where to take the blood?"

She looked puzzled, like maybe she needed to call security, so I told her.

"You take your thumb, and root around on the tail flukes, until you find a loose spot — and that's the vein."

She said, "Uh-huh," and got the needle.

She was testing my blood, to see if the cancer had come back.

When diagnosed with prostate cancer, there are multiple choices for how to treat it, and I had basically said "all of the above". Surgery? Check, had that, lots of bleeding, had to be re-admitted to the hospital when it wouldn't stop. Radiation? Check — thirty sessions, trapped naked inside this gigantic robot machine. Hormone therapy? Check — big injections, like harpoons flung across the room.

We were fighting the cancer the <u>second best way</u> we knew how. I say second best because I think the actual best way is some kind of stem cell/gene therapy, maybe one of those mentioned in this book.

The problem with fighting cancer is that it is hard to kill all of it. A vicious kind of stem cell is the reason: the cancer stem cell (CSC).

Most stem cells help the body, from embryonic stem cells which build us within the womb, to adult stem cells, which regrow skin after an injury.

But there are also Cancer Stem Cells (CSCs), which grow death.

There are multiple kinds of CSCs: leukemia cells took my sister Patty, breast cancer cells killed my mother, brain tumors stole took the life of Beau Biden, prostate cancer threatens me, and more.

Sometimes, after treatment, it may seem the cancer has gone away. But this could be temporary, as when my sister regained her strength after chemo. She looked beautiful at her wedding. But then the leukemia came back, and the cancer cells multiplied, and she died.

Stem cell pioneer Dr. Irving Weissman of Stanford has been fighting cancer stem cells for more than twenty years. In 2012, TIME Magazine said of his efforts:

"In a recent study, scientists reported that they successfully tested an antibody treatment that shrank human breast, ovary, colon, bladder, brain, liver and prostate tumors transplanted into mice."[1]

He had developed something bold and new, and it seemed to work — in mice.

---

[1] http://healthland.time.com/2012/03/28/a-single-antibody-to-treat-multiple-cancers/

Six years later, in a small clinical trial with humans, the progress continued.

As you recall, the first phase of clinical trials is just safety — to not make the condition worse. "First, do no harm," the Hippocratic Oath demands.

Not only did the patients in the test not get worse, some of them actually got better.

"Half of the 22 people enrolled in Phase 1 of the trial had a positive response to the therapy, and about one-third went into complete remission from their cancer."[2]

Dr. Weissman and I spoke on the phone recently.

He talks rapidly and uses mostly big words. It was a struggle to take notes.

Some things irritate him a lot, like:

"Some doctors even now deny the existence of cancer stem cells. They insist on doing what they did before, even if the patients died!"

The reason for his frustration was clear when he mentioned a previous test with standard cancer treatment in which every patient died.

But as I understand his explanation of the deadly process:

First, the cancer stem cell develops a bunch of bad cells. Some of these the body kills, chemotherapy or radiation may kill still more — but not the cancer stem cells.

The body's defense system cannot find them.

Think of cancer stem cells as microscopic monsters with markers on their backs. These markers (a form of protein) identify the cancer cells as bad, producing what Dr. Weissman calls "EAT ME" signs. If the body's immune system does find them, it will send fighting cells (macrophages) to absorb and destroy them. Sounds good, right?

But — *cancer stem cells can hide*. They grow a little "cloak of invisibility", a coating known as CD47, to cover the "EAT ME" surface markers. The immune system now does not recognize the threat, and ignores the cancer stem cells.

Weissman and his team developed an antibody, named HU5F9, which is designed to peel back the "cloak" so the cancer stem cells can be found and killed.

---

[2] https://med.stanford.edu/news/all-news/2018/10/anti-cd47-cancer-therapy-safe-shows-promise-in-small-trial.html

A possible way to fight all forms of cancer? This would seem to guarantee grants.

Unfortunately, Dr. Weissman says:

"One of the greatest problems in research today is the over-conservative nature of research funding. When I first applied for a grant to fight cancer stem cells, I was turned down by the National Institutes of Health (NIH) and other national cancer organizations."

"But the California Institute for Regenerative Medicine (CIRM) has a review panel of experts, and between their approval and the larger body of the Independent Citizens Oversight Committee, (ICOC, the program's public board of directors) my work was funded."

"Today, we are deep into clinical trials. I cannot say much until they are complete, but it is safe to say we are making therapeutic advances."

"And I can tell you this: Every time I present this, someone will say: you know you can't cure cancer — to which I have to say, **bullshit, we can't cure cancer!**"

— Irv Weissman, personal communication.

Making therapeutic advances...to me that sounds like something is getting fixed.

Yesterday as this is written, we heard from the co-founder of a company based on the Weissman battle against cancer.

His name is Mark Chao, he has worked with Dr. Weissman many years, and the company is called, appropriately, FortySeven.

Dr. Chao spoke at the May 23, 2019 meeting of the board that oversees CIRM.

He shared some enthusiasm and good news:

"We have a first-in-class anti-CD47 antibody called Hu5F9-G4 which has shown activity in patients enrolled on clinical trials with several types of cancers including non-Hodgkin's lymphoma, acute myeloid leukemia, myelodysplastic syndrome, and ovarian cancer. While the clinical trial experience continues to evolve and is still in early days, we in the company are working hard to try and bring this novel therapy to patients." — Mark Chao, personal communication.

We can also learn from CIRM's Kevin McCormack (KM), who interviewed one of the patients involved in the clinical trials.

KM: "In March of 2015, Tom Howing was diagnosed with stage 4 cancer. Over the next 18 months, he underwent two rounds of surgery and chemotherapy. Each time the treatments held the cancer at bay for a while.

"But each time the cancer returned. Tom was running out of options and hope when he heard about a CIRM-funded clinical trial using a new approach."

Tom Howing: "…After the cancer came back again they recommended I try this CD47 clinical trial. I said absolutely, let's give it a spin…

"Whenever you are dealing with a Phase 1 clinical trial (the earliest stage when the goal is first to make sure it is safe) there are lots of unknowns."

"Scans and blood tests came back showing that the cancer appears to be held in check. My energy level is fantastic. The treatment that I had is so much less aggressive than chemo; my quality of life is just outstanding."

"The most important thing I would say is, I want people to know there is always hope and to stay positive."[3]

As for my own blood tests, they came back negative: no cancer regrowth.

But in six months, there will be another visit to another nurse, and I will tell him/her my story, of how to take blood from a dolphin's tail.

---

[3] https://blog.cirm.ca.gov/2018/01/04/how-tom-howing-turned-to-stem-cells-to-battle-back-against-a-deadly-cancer/

25 Two Bulldogs

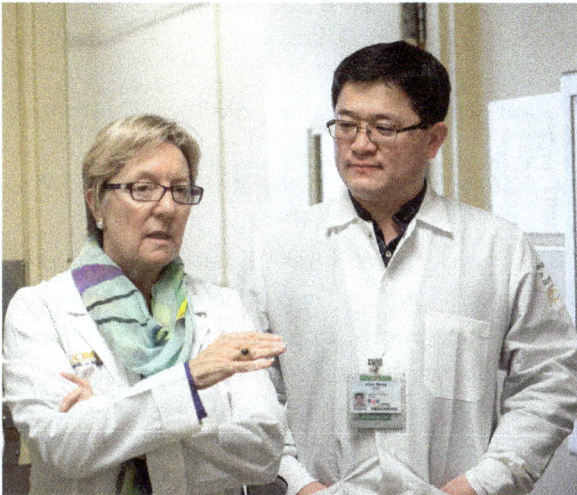

At UC Davis, Diana Farmer and Aijun Wang are battling spina bifida. (UC Davis photo.)

Imagine a blood blister, big as your fist, bulging out of a baby's back. This is Spina Bifida (SB).

There are several kinds of SB, and the worst is myelomeningocele (my-uh-low-ma-nin-jo-seal), in which "...a sac of fluid comes through the baby's back. Part of the spinal cord and nerves are in this sac and are damaged..."[1]

Kevin McCormack, director of communications for the California Institute for Regenerative Medicine (CIRM), says of SB:

---

[1] https://www.cdc.gov/ncbddd/spinabifida/facts.html

"Every day in America, four children are born with spina bifida, (which may result in) lifelong paralysis...The current standard of care is surgery, but even this leaves almost 60% of children unable to walk independently... Spina bifida involves many operations, many stays in the hospital..."[2]

But the California stem cell agency is fighting back. More from McCormack:

"Diana Farmer, MD and Aijun Wang, PhD, both at UC Davis, will use mesenchymal stem cells, taken from a donor placenta, and place them over the injury site in the womb. Tests in animals showed this approach was able to repair the defect, and prevent paralysis.

"Drs. Farmer and Wang have been working on this approach for more than ten years..."

The CIRM grant which made this possible? Five and a half million dollars.

Think about that for a moment. What if you had to personally raise five million?

For more than two decades, I have been involved in the struggle to raise money for research funding. Ever since my son Roman was paralyzed in a college football accident (September 10th, 1994) I have helped to raise funds: first for his rehabilitation costs, then for the Christopher and Dana Reeve Foundation, then a state law which raised $15 million for paralysis research — then for Proposition 71, Bob Klein's great citizen's initiative which raised $3 billion...

Individual charity or governmental funding — which is better? I have done both — and there is no comparison.

Example: My junior high students and I put on a play to raise funds for spinal cord injury research, led by paralyzed Christopher "Superman" Reeve. It took all year to write and produce a play about Mexican-American revolutionary Juan Cortina. It raised the magnificent sum of $4,000, in crumpled singles and coins, admission to our play and the sale of refreshments, cookies and soda. Two ladies from the Parent Teachers Association (PTA) sat on the stage and counted it in front of everyone, then sent a check to Christopher Reeve...the money raised was couch change compared to the need.

Of course, many charitable efforts are more successful than that, and I respect all efforts to seek cure, and ease suffering.

---

[2] https://blog.cirm.ca.gov/author/kmccormack2014/

If you visit the website of the Spina Bifida Association, you can see the hard work of families all across the country. Their efforts deserve honor.

But, if I understand correctly, their total annual budget (2016 is the latest year I could find) was about two million ($2,397,676).

It is endlessly difficult, raising funds. Every nickel is earned by somebody giving up their time, harassing friends and neighbors, for a few dollars in donation, and the need is so great.

Walk-N-Roll[3] appears to be the largest Spina Bifida fundraiser. If you go to their website, consider chipping in a couple bucks. I did ($25 dollars), which is not much, but that is how it is done. Everybody gives a little, and it adds up.

But California voters fought incurable disease — *as a state*. And that's why the California Institute for Regenerative Medicine (CIRM) was able to provide that major grant of $5.6 million, to fight spina bifida.

That money makes possible the work of folks from UC Davis: Diana Farmer, surgeon; Aijun Wang, stem cell expert; and Dori Borjesson, animal authority.

The operation will be done inside the mother, implanting a biodegradable mesh screen covered with stem cells (mesenchymals) derived from placenta. This, it is hoped, will heal what is wrong.

Animals get this terrible disease. If we can make them better, we not only ease their suffering, but also learn more on how to do it for humans. Dr. Borjesson is director of the U.C. Davis Veterinary Institute for Regenerative Cures. **The technique has been tried on two bulldogs, Spanky and Darla, who both walked normally after the operation.** It has also worked on lab rats, sheep, and cats.[4]

Might there be a way to use this technique (if it works) on people paralyzed by other conditions — like my son? A step forward for one condition can sometimes help another. At this stage, we do not know.

If all goes well, Dr. Farmer's effort (she is the Primary Investigator) will be approved for clinical trials by the FDA. After that, we will know more.

May their research be supported, until the need for it goes away...

---

[3] http://tinyurl.com/y83omvq9

[4] https://www.ucdavis.edu/news/stem-cell-treatment-children-spina-bifida-helps-dogs-first/

26 Other People's Pain: Fighting Bowel Disease

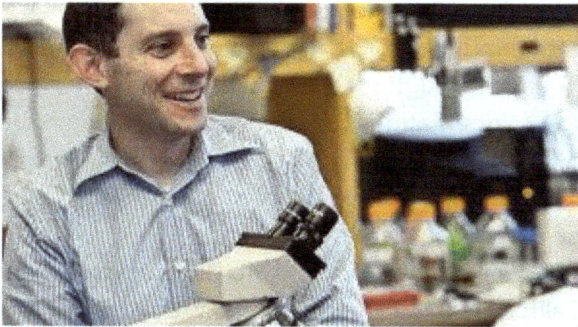

California funds Dr. Ophir Klein as he uses "organoids" to attack inflammatory bowel disease. (CIRM photo.)

No one ever understands another person's pain.

Before my son's accident, I thought people in wheelchairs just sat down a lot. I had no idea of the endless quiet agonies and frustrations they endure.

But people can notice a wheelchair, and at least understand something is wrong. What if a person looked utterly healthy, young, slender, beautiful — how could we know what was happening inside?

Inflammatory Bowel Disease (IBD) is a vicious condition. And yet, nothing shows, just that they may need the restroom (literally) a couple dozen times a day.

If an IBD sufferer gets tired of explaining their situation and says, "Oh, it is just stomach trouble," that is their right, of course.

But to suffer in silence may trivialize a deadly condition. Some body processes are embarrassing to discuss: so much so that in some cases (like colorectal cancer) people may literally die before they a talk to a doctor about it. By waiting too long, the disease may worsen beyond repair.

Also, there is the matter of research funding. If patients suffer in silence, decision-makers may give money to a noisier constituency. No noise, no funding — no cures.

How bad is Inflammatory Bowel Disease? (Note: there are a number of diseases called IBD, commonly Crohn's Disease and ulcerative colitis. My information is primarily based on Crohn's.)

If the food goes through the body too quickly, the nutrients of the food are not absorbed. The person loses weight in an unhealthy way: muscles wasting, slow starvation.

IBD is an autoimmune disorder; the body fights itself. Ulcers bleed — actual open wounds throughout the intestine. So much blood can be lost that the person faints. Fatigue is a constant companion. Just standing up can be too much. Arthritis can be so severe you cannot lift your arms to comb your hair. Diarrhea is common, as are stomach aches, like being punched in the gut. It can also lead to cancer.[1]

One and a half million Americans suffer Inflammatory Bowel Disease.

But the California Institute for Regenerative Medicine (CIRM) is fighting back, led by scientists like Ophir Klein (no relation to Bob) — and inspired by patient advocates like Rachel Bonner.

I heard Ms. Bonner, then 16, speaking before a meeting of the ICOC (Independent Citizens Oversight Committee) California's state stem cell board.

At age 8, Ms. Bonner went to the hospital with hip pain. At first, the doctors thought it was something in the bones. It would take six years and a recorded 241 medical treatments (!), as hospital staff struggled to figure out what was wrong, and what could be done to help.

You can hear her speaking.[2]

California is supporting Dr. Klein with financial grants to attack IBD.

His weapon of choice? Organoids. Remember the old movie, "THE BLOB"? It was Steve McQueen's first starring role, as a teenager who leads the fight against a jelly-like monster from outer space. The blob grew and grew until it ate a movie theater. The scientists defeated it

---

[1] https://www.crohnsandcolitis.com/crohns/disease-symptoms
[2] https://www.cirm.ca.gov/our-progress/video/living-ibd-inflammatory-bowel-disease-rachel-bonner-hope-crohns

finally in a very intelligent way, scooping it up and dropping it off in the North Pole, where it was too cold to grow.

But what if we had a blob as a friend, and could regrow wounded parts of the body? Organoids are mini-blobs, tiny chunks of living tissue. Implanted in the intestine, it is hoped, they will grow new and healthy tissue to heal the wounds. (Did you ever notice how similar the words "heal" and "healthy" are? When the body can heal itself, it is healthy.)

"The objective of the project is to (use) human embryonic and adult stem cells in the treatment of inflammatory bowel disease ..."[3]

Because of Proposition 71, the patient advocate-led initiative, there was money for Dr. Klein to fight the disease with science. But what if there had been no Prop 71, no patient advocate action?

The struggle is hard, and we, patients and families, must lead it. We are the 46%, the people with one or more chronic diseases.

Rachel Bonner organized a group and a website.[4]

At that website, are words that show what we must do.

"For the first few years of my Crohn's lifestyle, I was embarrassed to talk about it. I didn't want to share my struggle, the hardships of having Crohn's. But then I realized that talking about it is important, because for every one person you tell, they go and tell three more people. And that means four more people who know about (IBD) and the conversation will just keep spreading."

"As I go into my tenth year with this Crohn's lifestyle, I think about the doctor who told me when I was (a child) that I would only have eight or ten more years to live. I think about what people told me I couldn't do because of Crohn's."

"But the best part about having a stubborn resilient disease is that it teaches you to be stubborn and resilient yourself. You have to tell yourself that for every 'can't' ... you will show them a million reasons why you can and will live your life." — Rachel A. Bonner.

---

[3] https://www.cirm.ca.gov/our-progress/awards/stem-cell-therapy-inflammatory-bowel-disease
[4] http://www.hopeforcrohns.org/

27 Building Bone Density

Ying Zhu of Ankasa Regenerative Therapeutics seeks to alleviate bone-weakening diseases like osteonecrosis. (Ankasa photo.)

Late for my doctor's appointment, I parked the car and took off running, fast as my senior-seventy legs could shuffle. Out of the parking lot, up to the walkway. Almost there — and then the toe of my right shoe found a ripple in the concrete.

It is surprising how much time there is to think during an emergency. As I fell, I put my forearms in front of me. Also, I ducked my head to take the blow on the thickest part of the skull.

Even so, the impact was stunning. As I slammed against the concrete, I thought I felt my brain shake. I lay still for a second, taking inventory.

Skin scrapes, a headache, maybe a mild concussion?

The sounds of onlookers reminded me I was still late. I got up and ran off.

What a marvelous thing is our skeleton, that internal set of body armor! After a potentially serious fall, I was essentially unharmed.

But what if I had a disease which thinned the bones?

Osteo-**porosis** and osteo-**necrosis** are common bone-weakening diseases.

Sadly, there are cases (rare), where drugs to fight the first bone disease may bring on the second. Medication to treat osteoporosis may result in osteonecrosis where "the jaw rots and thighbones snap."[1]

Imagine being the doctor with such a choice! How would you tell a patient about a medication which had a slim but real chance of such horror-show side effects? You would of course describe what could happen. But what if your patient was then so scared of the side effects of "necrosis" that he/she would not take the medication — and thereby suffered broken bones? Hip fractures can be fatal ...

Read this devastating paragraph ...

"Estimates of (Americans) with osteoporosis range from ten to twenty-five million, with 75% being women...(Of these) between 1.5 to 2 million a year (get) fractures which may land otherwise healthy elderly people in nursing homes for the remainder of their lives...Osteoporosis costs the nation an estimated $19 billion a year."[2]

The California stem cell program is challenging both forms of bone disease.

For this article, let's concentrate on osteo-necrosis, which means "bone-death."

Ying Zhu is a CIRM-funded scientist working for Ankasa Regenerative Therapeutics, in La Jolla, California.

"Osteonecrosis is a painful, progressive disease for which there is no treatment, (except) replacing the dead bone with a metal implant."

Dr. Zhu's approach appears to be (to my non-scientist's eye) to take a piece of the patient's bone (called an autograft) and place it in the wound.

The problem? "Autografts contain stem cells. In young patients, these stem cells ... give rise to new bone, but in older patients, autografts are ineffective ..."

---

[1] https://www.nytimes.com/2016/06/02/health/osteoporosis-drugs-bones.html
[2] https://www.cirm.ca.gov/our-progress/osteoporosis-bone-and-cartilage-disease-fact-sheet

A potential answer? According to Zhu, her new therapeutic "re-activates the stem cells in an older person's autograft... the resulting material (called ART 1001) generates more (bone) cells, and engrafts better..."[3]

Dr. Zhu and Ankasa Regenerative Therapeutics are currently submitting the all-important IND (Investigational New Drug) submission to the FDA. — Ying Zhu, personal communication.

On behalf of all who intend to grow old in good health, I wish them every success!

---

[3] https://www.cirm.ca.gov/our-progress/awards/autologous-somatic-stem-cell-therapy-treatment-osteonecrosis

28 Money, Hope, and Huntington's

Frances Saldana, patient advocate fighting to defeat Huntington's disease, which took the lives of her family members. (News.uci.edu photo.)

Imagine coming home from work to find one of your family members — changing into someone else?

Sound like one of those 1950s monster movies: "I Married a Monster from Outer Space"? Unfortunately, this is grim reality.

This year, one in ten thousand Americans will be diagnosed with Huntington's Disease (HD), one of the worst conditions imaginable.

Frances Saldana has lived the tragedy of Huntington's.

As you recall if you read my previous book, "CALIFORNIA CURES," her husband Hector had the condition, and died struck down by a car as he walked awkwardly across a street. Their daughter Marie, institutionalized with HD, became friends with the woman whose car struck Hector. Mother Frances did not want to interfere with her daughter's happiness, and never revealed the secret; Marie died not knowing her friend had accidentally killed her father.

Hector died, and Marie died, and Margie died.

And as this is written, Frances' son Michael is in a coma, from which he is not expected to return.

Talk to any Huntington's family, and you will hear a similar tragedy. It is incurable, and fatal.

But Frances is fighting the disease, raising money toward its defeat.

She has raised as much as $70,000 at one event, all of which goes to the research. But these are thousands, where millions are needed, which is why the California Institute for Regenerative Medicine (CIRM) is so vital.

And CIRM is running out of money...

Crucially, Ms. Saldana got a job at UC Irvine, one of the most respected medical institutions in the world. There she met Dr. Leslie Thompson, who has worked 30 years to find a cure for Huntington's.

Dr. Thompson is using a variety of stem cell approaches to "provide neuroprotection to the brain (as well as) the possibility of cell replacement." In other words, she is trying to both prevent further damage by Huntington's, as well as hopefully repairing the damage already done.

Using skin cells taken from people with Huntington's Disease (HD), and a cocktail of chemicals, Dr. Thompson and her colleagues made an HD stem cell line. With this "disease in a dish," it is possible to try different medicines on the cells and see what works, without endangering a human.

Her second approach involved embryonic stem cells, which would be changed (differentiated) into nerve stem cells. If those healthy cells are transplanted, they might protect patient nerves from being destroyed, perhaps by blocking the protein that causes HD harm.

This might help defeat not only HD but also other nerve-damage diseases like ALS (for which her colleague Clive Svendsen just received FDA permission to begin human trials), Alzheimer's, or Parkinson's disease.

Dr. Thompson has gathered a team of experts: Mathew Blurton-Jones, Ed Monuki, Steven Cramer, and Neal Hermanowicz of UCI; Michael Levine and Marie-Francois Chesselet of UCLA; Clive Svendsen of Cedars-Sinai Medical Center; and Gerhard Bauer of UC Davis.

The group is partnering with Terumo BCT, a global biotech company that employs a system for expediting the growth of stem cells. BioTime, Inc., (begun by stem cell pioneer, Mike West)...is providing the...cells. These will be developed into neural cells at the Good Manufacturing Practices (GMP) facility at UC Davis.

The work would be headquartered at the magnificent Sue and Bill Gross Stem Cell Research Center at UC Irvine.

But for the fight to be won, funding must be found.

For me, it all came together on January 30th, 2019. First, Leslie Thompson would speak, following which I would be allowed three minutes to make a public comment on Huntington's grant, CLIN 1-10953.

Dr. Thompson said (she shared her notes with me afterward):

"HD is an unrelenting and devastating disease that strikes in the prime of life, or younger...Although prospects for treatment are more promising than ever, nothing is yet available to change the course of disease. And it may be that we need combinations of different treatments for HD."

"We propose to use a human stem cell-based transplantation approach. There has been extensive investment and guidance from CIRM. We have consulted with the FDA, clinicians and scientists from around the world, to leverage ongoing collaborations in the field and learn from previous experiences to develop best practices for moving forward. We propose to carry out pivotal safety studies and prepare for clinical trials."

"Will this slow progression or improve outcome?"

"We know that (a) there is neuroprotection to keep brain cells healthy and restore factors that are lost, (b) transplants reduce toxic form of mHTT — the poison that kills the cell and (c) there are robust positive effects in 3 HD mouse models. As a component of the Clin1, we will delve deeper into the mechanisms that underlie the strong positive benefit."

"I have been in the HD field for 30 years and have built relationships and seen the devastation of the disease to families. I am committed to helping the families and believe this path can have benefit. I hope you will approve funding." — Leslie Thompson, personal communication."

After Dr. Thompson, I would have a turn, three minutes: 180 seconds to speak on behalf of friends with Huntington's.

Some people like to "wing it," talking off the top of their heads, saying whatever comes to mind.

But for me, when it is this important, I like to have every word written out:

"When CIRM takes on a challenge, it is because people are suffering — but maybe there is a way to ease the agony. Seldom is this more clear than in the battle against Huntington's. The suffering is unquestionable, and there is no cure: not yet."

"The disease itself is terrible: like that sentence from Shakespeare, 'Those whom the gods would destroy, they first make mad'."

"Huntington's does that. Bad enough that it slowly kills the sufferer physically — 15–20 years of suffering before death — but it affects their minds as well. Making them sometimes foul-tempered, or removing their good judgment. Huntington's may make the sufferers literally insane. The family constantly has to remind themselves, this is the disease, not him — and they may go through this for decades."

"But is there hope for this particular approach: CLIN1-10953?"

"Will Brain-Derived Neurotrophic Factors — sort of nerve fertilizers — help in the fight against Huntington's?"

"Can the scientists decrease the levels of something called mutant HTT protein — which may in fact cause the condition?"

"Lastly, will the newly added neural stem cells — healthy ones — overcome the sickness of Huntington's disease?"

One thing we know already:

"When you think of people qualified to fight Huntington's, here they are: top-notch folks, champions in their field. Some have studied Huntington's for more than 30 years; they examined the possibilities; this is what they believe will help."

"I thank the board for its attention. Please give the experts the funding they need to challenge this vile and hateful condition."

The vote was taken; the answer was yes — on this day, there was enough money.[1]

---

[1] https://www.cirm.ca.gov/our-progress/disease-information/huntingtons-disease-fact-sheet

29 A Better Rat?

"In China, this is the year of the rat," said Qilong Ying. (Ying Lab photo.)

When I told my wife Gloria I was writing an article about rats, she had several comments, including: "Oo, ugh!" and also "That's disgusting!"

Obviously, we have problems with rats, such as when they chew through electrical wires, which may cause a short circuit and burn down the house. Also, they are blamed for carrying diseased fleas in their ears and spreading the Black Plague, which in 1340 killed half of China and one-third of Europe — but this is not certain. The plague may in fact have been transmitted by human-carried parasites.

But there are positive aspects to rats as well. For instance: "...a rat paired with another that has a disability...will be very kind to the other rat. Usually, help is offered with food, cleaning, and general care."[1]

---

[1] https://www.amazon.com/Rat-Owners-Guide-Happy-Healthy/dp/1620457393

Above all, anyone who has ever been sick owes a debt to rats, specifically the Norway rat with the spectacular name, *Rattus norvegicus domesticus*, found in labs around the world.

I first realized its importance on March 1, 2002, when I held in my hand a rat which had been paralyzed, but then recovered the use of its limbs.

The rat's name was Fighter, and she had been given a derivative of embryonic stem cells, which restored function to her limbs. (This was the famous stem cell therapy begun by Hans Keirstead with a Roman Reed grant, developed by Geron, and later by CIRM and Asterias, Inc., which later benefited humans.)

As I felt the tiny muscles struggling to be free, it was like touching tomorrow — while my paralyzed son, Roman Reed, sat in his wheelchair just a few feet away.

Was it different working with rats instead of mice? I had heard that the far smaller lab mice were more "bitey" than rats.

Wanting to know more about the possibilities of a "better rat", I went to the CIRM website (www.cirm.ca.gov), hunted up the "Tools and Technology III" section, and the following complicated phrase:

"Embryonic stem cell-based generation of rat models for assessing human cellular therapies."[2]

Hmm. With science writing, it always takes me a couple of readings to know what they were talking about. But I recognized some of the words, so that was a start.

"Stemcells...rat models...human therapies..."

I called up Dr. Qilong Ying, Principle Investigator (PI) of the study.

As he began to talk, I felt a "click" of recognition, as if, like pieces of a puzzle, facts were fitting together.

It reminded me of Jacques Cousteau, the great underwater explorer, when he tried to invent a way to breathe underwater. He had the compressed air tank, and a mouthpiece that would release air — but it came in a rush, not normal breathing.

So he visited his friend, race car mechanic Emil Gagnan, and told him: "I need something that will give me air, but only when I inhale," — and Gagnan said: "Like that?" and pointed to a metal contraption on a nearby table.

---

[2] https://www.cirm.ca.gov/our-progress/awards/embryonic-stem-cell-based-generation-rat-models-assessing-human-cellular

It was something invented for cars. But by adding it to what Cousteau already had, the Cousteau–Gagnan SCUBA (Self Contained Underwater Breathing Apparatus) gear was born — and the ocean could now be explored.[3]

Qi-Long Ying's contribution to science may also be a piece of the puzzle of cure...

A long-term collaboration with Dr. Austin Smith centered on an attempt to do with rats what had been done with mice.

In 2007, the Nobel Prize in Medicine had been won by Dr. Martin Evans, Mario Capecchi, and Oliver Smithies. Working independently, they developed "knock-out" and "knock-in" mice, meaning to take out a gene, or put one in.[4]

But could they do the same with rats?

"We and others worked very, very hard, but got nowhere," said Dr. Evans.[5]

Why was this important?

Many human diseases cannot be mimicked in the mouse — but might be in the rat. This is true for several reasons: the rat is about ten times larger, its internal workings are closer to those of a human, and the rat is considered several million years closer (in evolutionary terms) to humans than the mouse.

In 2008 ("in China, that is the year of the rat," noted Dr. Ying in our conversation) he received the first of three grants from CIRM.

"We proposed to use the classical embryonic stem cell-based gene-targeting technology to generate rat models mimicking human heart failure, diabetes and neurodegenerative diseases..."[6] How did he do?

In 2010, Science Magazine honored him with inclusion in their "Top 10 breakthroughs" for using embryonic stem cell-based gene targeting to produce the world's first knockout rats, modified to lack one or more genes.

And in 2016, he and Dr. Smith received the McEwen Award for Innovation, the highest honor bestowed by the International Society for Stem Cell Research (ISSCR).[7]

---

[3] https://en.wikipedia.org/wiki/Aqua-Lung
[4] https://www.ncbi.nlm.nih.gov/pmc/articles/PMC2018769/
[5] https://www.nature.com/stemcells/2009/0901/full/stemcells.2009.11.html
[6] https://www.cirm.ca.gov/our-progress/people/qilong-ying
[7] https://www.eurekalert.org/pub_releases/2016-03/uosc-sqy031516.php

Using knowledge learned from the new (and more relevant) rat, it may be possible to expand stem cells directly inside the patient's own bone marrow. Stem cells derived in this fashion would be far less likely to be rejected by the patient. To paraphrase Abraham Lincoln, they would be "of the patient, by the patient and for the patient — shall not perish from the patient" — sorry!

Several of the rats generated in Ying's lab (to mimic human diseases) were so successful that they have been donated to the Rat Research Resource center so that other scientists can use them for their study.[8]

"Maybe in the future we will develop a cure for some diseases because of knowledge from using rat models," said Ying. "I think it's very possible. So we want researchers from USC and beyond to come and use this technology."[9]

And it all began with the humble rat...

---

[8] http://www.rrrc.us/ES_Cells_/
[9] https://yinglab.usc.edu/ying-lab-produces-first-conditional-and-inducible-rats/

30 "Tuesdays with Morrie": Battling ALS

Morrie Schwartz' battle with ALS inspired Mitch Albom's book, "Tuesdays with Morrie." (YouTube.com photo.)

"Tuesdays with Morrie" is the true story of the last days of Morrie Schwartz, punctuated by the visits of his student, author Mitch Albom.[1]

"Tuesdays" became a huge international favorite. To his credit, the author used the book proceeds to help with Schwartz's medical bills. After Morrie's passing, Albom went on to develop a radio show — which I got to visit!

It was a phone-in show, a good thing, saving me a drive to Detroit, Michigan.

Monday, June 11, was my wife's birthday, and she sat in the good chair in my study, while I endured the last nervous moments before the show.

You can listen in if you like, the url is below.[2]

---

[1] https://www.amazon.com/Tuesdays-Morrie-Greatest-Lesson-Anniversary/dp/076790592X

[2] https://www.mitchalbom.com/radio/

I was on at 3:35, Pacific Standard Time, for seven minutes.

My goal was to talk about the California stem cell program, the subject of my new book, "CALIFORNIA CURES: How the California Stem Cell Program Is Fighting <u>Your</u> Incurable Disease!"

The California Institute for Regenerative Medicine (CIRM)... It was hard to know where to begin describing it, because the program has done so much.

But re-reading "Tuesdays with Morrie" gave me focus.

Morrie Schwartz died of ALS, Amyotrophic Lateral Sclerosis, Lou Gehrig's disease, called motor neuron disease in many countries.

Every year, six thousand more Americans are diagnosed with ALS.[3]

It is a death sentence. They will live just two to five years after the disease is diagnosed. ALS is an auto-immune disorder, meaning the body attacks itself, killing its own motor nerves, primarily in the spinal cord. As the motor nerves die, messages from brain/spinal cord to muscles don't get through.

"When a healthy person wants to move a leg, for example, the brain sends a signal...to the motor neuron that enables the movement. Patients with ALS develop progressive paralysis..."[4]

The muscles degenerate; control of the body goes away. Eventually, the patient loses the ability to breathe.

We have lost far too many good people to ALS, like:

Lou Gehrig, the New York Yankee baseball player, for whom the condition is named.

Dwight Clark, the great San Francisco Forty-niner football player.

David Ames, patient advocate, who fought to pass Proposition 71.

The two sons of Dianne Winokur, Hugh and Douglas: Ms. Winokur serves on the board of directors for the California Institute of Regenerative Medicine.

Most recently, the great scientist Stephen Hawking, who somehow defied the odds and survived half a century past his diagnosis.

Before going on the show, I visited the website of the California stem cell program (www.cirm.ca.gov) to learn about the battle against ALS.

The California Institute for Regenerative Medicine (CIRM) fights chronic disease; that is its purpose, taking on diseases called "incurable,"

---

[3] http://www.alsa.org/news/media/quick-facts.html
[4] https://www.cedars-sinai.org/newsroom/cedars-sinai-receives-approval-to-test-novel-combined-stem-cell-and-gene-therapy-for-als-patients/

and step by step working towards their defeat. The scientists there are honest folk, not telling you something wonderful just to make you feel good temporarily. They work on hope, not hype.

How goes the fight for ALS cure? Hard, slow, but making progress. CIRM has spent $79.4 million to learn about the disease and the methods of fighting it.[5]

Right now, CIRM is helping fund two ALS human trials: one at Cedar-Sinai Regenerative Medicine Institute, and the other at a private corporation, BrainStorm Cell Therapeutics. They are funded by CIRM to the tune of roughly $16 million each. (Cedars-Sinai, $16,168,464; Brainstorm, Inc., $15,912,390.)

This is serious funding. By way of comparison, another California research program, the Roman Reed Spinal Cord Injury Research Act, received only about $15 million for its entire ten year program!

Cedars-Sinai, led by Clive Svendsen and James Baloh, is trying to boost the power of astrocytes, star-shaped cells which support the motor nerves. They will be boosted by a "nerve fertilizer" called Glial-Derived Neurotrophic Factor (GDNF).

A second clinical trial, by a private organization called Brainstorm Inc., uses mesenchymal stem cells from the patient's bone marrow. "These stem cells are then modified to boost their production of (several) neurotrophic factors... to support and protect (motor nerves)... destroyed by the disease."

Now in multiple sites across the country, the BrainStorm effort began as an international effort in Hackensack, New Jersey and Petach Tikva, Israel.

"The CIRM funding will enable the company to test this therapy, called NurOwn, in a phase 3 trial involving around 200 patients."

— ALS Fact Sheet, California's Stem Cell Agency.

Right now, the goal appears to be to slow the progression of the disease, to gain time, years of healthier life to spend with their families.

But that's not enough. We need a cure. That's why we have to continue the funding of CIRM, to renew the program, so the battle can go on — to victory.

Remember Morrie Schwartz!

---

[5] https://www.cirm.ca.gov/our-progress/disease-information/amyotrophic-lateral-sclerosis-als-fact-sheet

31 Punching at Parkinson's

"Just the normal thing, go in and have a couple holes drilled in your head and some wires put in," — newscast legend Ronn Owens describes Deep Brain Stimulation. (Ronn Owens Report photo.)

Years ago, I was interviewed by famed KGO Radio talk show host Ronn Owens and was impressed by his asking just the right questions. Interviewing is a subtle skill, and what seemed like casual conversation — warmth and encouragement spiced up with humor — was in fact a friendly interrogation. Afterward I felt as if my brain had been vacuumed.[1]

So, when my first book on stem cell research came out, naturally I thought of contacting the Bay Area's legendary journalist, only to find he was just recovering from surgery: he had had Deep Brain Stimulation (DBS) to ease the symptoms of his Parkinson's disease.

Asked what he was doing in the hospital, Owens replied:

"Just the normal thing, go in and have a couple holes drilled into your head and some wires put in."

DBS has been called a pace-maker operation for the brain. A battery device the size of a silver dollar is put under the skin near the collarbone,

---

[1] http://www.kgoradio.com/ronnowens/

with a wire going up through the neck into the brain; there it will produce mild shocks to counteract the "abnormal nerve signals that cause PD symptoms."

If the operation went well, which it apparently did, many of the problems of Parkinson's (tremors and shaking) would be considerably reduced, perhaps for 5–7 years. And, conveniently, if something better is found, the device can be removed.

If I had Parkinson's, I would definitely investigate Deep Brain Stimulation.

But while DBS may help with symptoms, it does not cure the disease.

If you visit the California Institute for Regenerative Medicine (CIRM) website, you will find about thirty different approaches (and $54 million spent on the effort) attempting to defeat Parkinson's.

This is as it should be. In America alone, PD afflicts one million people, an enormous burden to families. We need to try many different ways until we can find the cause, and can implement a cure.[2]

Remember Thomas Edison and the light bulb?

"In the period from 1878 to 1880 Edison and his associates worked on at least three thousand different theories to develop an efficient incandescent lamp."[3]

Edison sought a filament that would carry light inside the bulb. They finally settled on — carbonized bamboo. When that worked, the world changed.

But what if the funding had run out? America might still be fumbling in the dark.

Key Fact: stem cells help the body produce dopamine, the natural chemical which makes motion smooth and coordinated.

I spoke with three top Parkinson's researchers (all recipients of CIRM grants) recently.

Dr. Xianmin Zeng of the Buck Institute received a CIRM grant for "banking transplant-ready dopaminergic neurons..."[4]

That makes sense. If there was a "bank" of different kinds of healthy nerve cells ready to go, the right match could be quickly available for someone in need. This would save time searching — and prevent

---

[2] https://www.cirm.ca.gov/our-progress/disease-information/parkinsons-disease-fact-sheet

[3] https://www.fi.edu/history-resources/edisons-lightbulb

[4] https://www.cirm.ca.gov/our-progress/people/xianmin-zeng

the body from rejecting the transplant. Also, having a variety of stem cells could be useful as a disease model: to test new therapies and medications.

One of the most exciting projects comes from Dr. Xinnan Wang of Stanford, who is working primarily on mitochondria, the power source for the cell.[5]

Here is my layperson's translation of what she is doing.

Imagine three things: a tiny machine, a battery, and some glue. The machine is the nerve cell. The battery is the mitochondria. And the glue? A sticky substance called Miro, which keeps the mitochondria in the right place.

What happens when a battery wears out? You throw it away.

In a healthy body, worn-out mitochondria just dissolve.

But if there is too much Miro, the mitochondria sticks on the cell too long, and exudes toxic wastes — *which may be the cause of Parkinson's*.

Dr. Wang's theory is supported in a bizarre way.

Fruit fly larvae (they look like lice) crawl at a certain rate of speed. If the amount of Miro is increased around them, their crawling rate slows down. If the Miro is later reduced, the larvae will crawl at normal speed.

Here is Dr. Wang:

"Prolonged retention of Miro, and the consequences that ensue, may constitute a central component of Parkinson's Disease."

"… partial reduction of Miro levels… in both fruit fly larvae and human cells … rescues (the subject) from neurodegeneration." — Xinnan Wang, personal communication.

What does Parkinson's cost America? Individually, someone with Parkinson's Disease must come up with about $22,500 a year in medical costs, more than double the medical expenses of a healthy individual. ($10,000 annually).

As a country? Out of pocket costs are roughly $14.4 billion. Indirect costs, like time lost from work, add another $6.36 billion. Altogether, America spends more than $20 billion dollars a year on this terrible condition — without curing it![6]

Suffering? Since the condition is progressive, no matter how bad you feel today, it may be worse tomorrow.

---

[5] http://med.stanford.edu/xinnanwanglab.html
[6] http://www.silverbook.org/fact/in-addition-to-the-estimated-14-4-billion-a-year-in-costs-for-parkinsons-disease-to-the-nation-an-additional-6-3-bil/

Finally, Dr. Jeanne Loring is a special friend to advocates, always willing to attend another meeting, or answer an email. She is also a long-term scientist at the world-renowned Scripps Institute, so what she said first startled me:

"I am leaving Scripps," she said, "to devote all of my time to bringing the therapy to the clinic. It's unusual for a professor to leave academia, but it is the right thing for me to do."

"Aspen Neuroscience raised $6.5 million in just a couple of months and was launched about 6 months ago. All of my equipment, intellectual property, my grants, and my scientific staff have moved to Aspen. ... The equipment was bought with my CIRM funding for the lab training course and shared lab, and my scientists were all supported mainly with my CIRM grants..."

She regards Parkinson's disease as:

"an ideal target for cell replacement therapy. The problems with movement in PD is caused by loss of a specific cell type — dopamine neurons in a specific part of the brain..."

"We have developed a patient-specific therapy, using dopamine neurons from induced pluripotent stem (iPS) cells from individual patients...and we hope to obtain approval to start a clinical trial in 2020." — Jeanne Lorring, personal communication.

Dr. Lorring, Dr. Wang, and Dr. Zeng — with minds like these (and proper funding!) we have a chance to defeat the scourge of Parkinson's.

And Ronn Owens? His condition remains, but he is dealing with it. He is not only still on the air, but doing outstandingly, being recently inducted into the National Radio Hall of Fame, the highest honor a radio broadcaster can achieve.[7]

---

[7] https://en.wikipedia.org/wiki/Ronn_Owens#Personal

32 Arthritis Champions

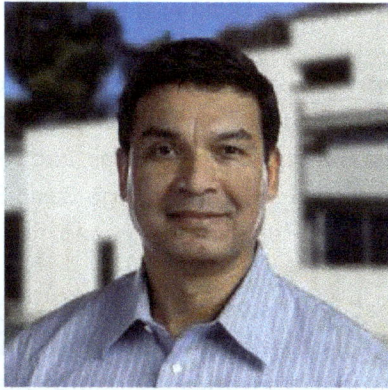

Arthritis scientist Darryl D'Lima won the Best Cartilage Science Award from the International Cartilage Repair Society in 2000, and has stood out ever since. (Scripps.edu photo.)

Gloria and I live in a two-story house — which is a problem. Ideally, Gloria uses the stairs just once every day: down in the morning, back up at night. Items which must be trucked up or down in-between are stacked on the steps for hubby.

An x-ray of her knees reveals why.

A healthy knee joint has an internal cushion of smooth white cartilage. Gloria's cushion is mostly gone. Her leg bones grind together at every weight-bearing step, and she suffers.

She is not alone. Arthritis affects one in ten (30 million!) people in America.[1]

Artificial knee joints exist: replacements can be made of metal and porcelain. But transplantation is painful and long. Recovery takes two–three

---

[1] https://www.cdc.gov/arthritis/basics/osteoarthritis.htm

months. For some, the process involves so much pain that they decline a second surgery on the other limb. Also, there is a distinct possibility that the surgery may not succeed, and need to be done over.

Might there not be a better way? Could the knee cushion be regrown — internally?

Visiting the CIRM website was a comfort as always; here were experts, fighting on our side.[2]

First, I found 4 completed projects attacking arthritis, for a total of $14,757,684. These were all done by champions: Peter Schultz of Scripps Research Institute, Dan Gazit of Cedars-Sinai Medical Center, Dennis Evseenko of UCLA; and Darryl D'Lima, director of orthopedic research at Scripps Clinic.

Each had a different approach, too complicated to go into right now.

But what interested me most was a fifth project, also by Darryl D'Lima, M.D., PhD., a leader in the field of joint repair and replacement.[3]

In 2000, Dr. D'Lima won the Best Cartilage Basic Science Award from the International Cartilage Repair Society, and he has stood out ever since.

D'Lima's project had the usual gigantic title: "Embryonic stem cell-derived chondroprogenitor cells to repair osteochondral defects" — scientists can't talk without big words! "Chondro" means bone and joint. "Chondro-progenitor cells" means the in-between cells that make those bones and joints.

The grant was a major one, for more than seven million ($7,660,2110), a good sign in itself. CIRM is careful with its grants.[4]

What was its approach?

Dr. D'Lima wants to do a knee implant with a gel structure, a scaffold, tough but biodegradable (it will slowly break down and be disposed of by the body) with start-up cells inside to grow the needed body tissues, replacing what was gone:

"...constructing scaffolds that are seeded with cells...programmed to (change) into bone and cartilage cells..."

It claimed "the unique advantage that the same material is universally applicable to all patients..." (within the targeted age group).

---

[2] https://www.cirm.ca.gov/our-progress/arthritis-fact-sheet

[3] https://www.cirm.ca.gov/our-progress/people/darryl-dlima

[4] https://www.cirm.ca.gov/our-progress/awards/embryonic-stem-cell-derived-chondroprogenitor-cells-repair-osteochondral-defects

It would not help Gloria or myself. Why?

The aim was "early osteoarthritis" for people under 55. Gloria and I both see seventy in the rear-view mirror. But the technique might be "tweakable", later on, and apply to oldsters like us.

Why choose the early form of the condition?

"For younger patients with severe arthritis or impending arthritis, there is no treatment that can prevent, cure, or even slow the progression of this disease."

For Gloria and me to have the condition is bad enough — but for young people to suffer almost their entire lives? That would be terrible indeed.

Instead, they might end up with what feels and acts like a new knee…

How is the new project working?

"Major milestones for year 1 set by CIRM were met. A cell manufacturing and quality control process was established to generate cell banks that would be safe and suitable for clinical work. The cells passed all established laboratory criteria to measure potency in repairing tissue and treating arthritis…The first pre-clinical experiments were initiated."

"The funding provided by CIRM is essential to the development and support of the research we are doing with regard to tissue regeneration at Scripps," said Dr. D'Lima. "With this grant we plan to continue our progress in this field and move toward clinical trials…."[5]

If they knew about his progress, I am sure thirty million suffering people — and their families — would wish Darryl D'Lima all success.

[5] https://www.scripps.org/news_items/5215-scripps-health-awarded-7-6-million-grant-from-california-institute-for-regenerative-medicine

33 Marching for Science?

Advocates must rally, to share energy and spread the word…. (Americans for Cures photo.)

I was trying very hard NOT to go to the April 14, March for Science, in Oakland, California.[1]

I had no real excuse to miss the event. I had already helped Roman (my paralyzed son) with his medical needs in the morning, plus my stem cell writing was pretty much done for the day, though of course there was always studying to do. But basically I had no conflicting appointments

---

[1] https://www.marchforsciencesf.com/

until 5:00, at which time my wife Gloria intended to drag me out to dinner with one of her numerous groups, grumble, grumble...

It was Saturday. I leaned back in my favorite chair, closed my eyes...

But the longer I lay there, the more awake I got. My legs twitched without my will, as if there was music playing nearby. And somehow, despite all my attempts to nap, I found myself vertical, putting on a favorite RIDE FOR ALS t-shirt, walking down the steps, out the door, into the car — and heading for the Bay Area Rapid Transit (BART) station.

After updating a "clipper" card to ride, I scrambled up the steps of California's beautiful mass transit system.

I was not entirely sure where I was going. One crowd of young people were gathered around a woman with a t-shirt labeled "I AM A GENIUS!" — I asked her if she was going to the March for Science rally? No, she was not. I told her that since her shirt said she was a genius I was sure she supported stem cell research!

Disembarking in Oakland, I asked which way was the Lake Merritt amphitheater, and walked a half mile or so, interacting with local folk. If they showed interest in science, naturally they got my twenty-second pitch for stem cell therapies.

Example: "I'm looking for the Lake Merritt Amphitheater, they're having a rally for science there?"

Pause for response: "I think Lake Merritt is that way"...

"Oh, great, I support stem cell research, and as you know, Oakland is the headquarters of the California stem cell program!"

If they showed signs of interest (i.e., not leaving), I told them about the girl in the plastic bubble whose life was saved by stem cell therapy, or my son Roman Reed's fight against paralysis; stem cell mini-messages, important for them to hear.

Closer to Lake Merritt, pro-science signs increased. A woman and her two sons had T-shirts about climate change — another block and turn right — and there it was.

When you see a rally on TV, it can seem a little threatening, as if violence was about to erupt, and you might be arrested or tasered, for speaking too loud.

But this was just friendly people, working together for the common good.

The Lake Merritt Amphitheatre is an open-air arena, a grassy inclined slope before a small stage. Local talent made music, and when they stopped a young lady speechified about global climate change.

If I had thought about the event earlier, I might have asked to speak myself, five minutes on the California stem cell research program. But I hadn't, so I didn't.

The big event was a 5-mile march around Lake Merritt. That was scheduled to start at 2:30... Let's see, five miles at my walking pace...I calculated my chances of survival if I showed up late for Gloria's dinner event. Hmm. Probably not.

What I liked was all the little booths set up everywhere, staffed by people like me.

Advocates recognize each other. We all have something important to say, and are generally seeking some way to share it.

It is like if you have the second most adorable grand-daughter. (I have the first). We would both let the other talk for a moment, while waiting for our turn.

And they absolutely needed to know Oakland's best-kept secret.

"Do you see that building over there, the tall one? On the 16th floor is the California stem cell research program — it began as a citizen action (advocates love that part) — and we hope it will be renewed in 2020!"

Then I would listen to their pitch, earthquake preparedness or gender equity or combating homelessness — sign a petition, take some literature, swap advocate cards.

This was important stuff.

Washington has too many members of what I call the Anti-Science Society (or A.S.S. for short).

One year ago, in various sites across the world, a million people gathered in support of science. In the Bay Area, perhaps 70,000 supporters rallied.

Today looked about 2,000, according to one of the organizers. It was — sustainable.

Everybody here was a pro-science advocate.

Some of the folks I talked to were:

An educator and science rapper, Glenn Wolkenfeld.

Dr. Shaye Wolf, of the Center for Biological Diversity.

Daniel Hilsinger of March for our Health.

Dr. Jonathan Foley, of the California Academy of sciences.

Samuel Gatachew of Youth Speaks.

Dr. Mayra Padilla, Dean of Institutional Equity at Contra Costa College.

Dr. Jennifer R. Cohen, American Association for the Advancement of Science policy Fellow.

Amy Hines-Shaikh, Healthy California —

And lots more.

I walked around for a couple hours, making friends for stem cell research.

Then I got tired, and went home.

Are such rallies important? You bet! Pro-science people need to know the strength we have, because — *we are overwhelmingly in the majority.* Let me prove that.

What percentage of Americans are against cuts in medical research?

Eighty-seven per cent.

Let me say that again. According to a recent national poll, 87% of Americans are against cuts in medical research. 87%? That is almost nine out of ten — it is hard to find anything agreed on by almost nine in ten Americans![2]

We are millions... All we need to win is to recognize our numbers, share our messages, vote appropriately, and (every so often) get together and enjoy the company of others like ourselves.

Want more information? Here is a useful guide to science advocacy.

Want to know about future Marches for Science?

Contact Holly Cordero, Communications Director for the March. (650-679-4006) press@marchforsciencesf.com

---

[2] https://poll.qu.edu/national/release-detail?ReleaseID=2444

34   Blood, Blood, Blood!

Tannistha Reya wants to make replacement blood cells, to be available for emergencies....
(Eurekalert.org photo.)

As a child with asthma–bronchitis I spent entirely too much time in hospital. I swallowed vile medicines, some which burned like liquid fire, others that tried to choke me as they went down.

There was also one huge mean nurse: the opposite of the gentle and kind Florence Nightingale — or maybe my nurse was just overworked, underpaid and had a backache from lifting patients all day — I only know I feared her, and with reason.

Once she packed me in ice chips to bring my fever down. She warned me, before she left: "*Now don't you move.* If you do, that will melt the ice, and I'll have to come back and do you again."

She yanked the oxygen tent's plastic curtains together, and I was alone.

For a while it was mildly interesting watching the steam of my breath congeal on the plastic curtain, beading into water droplets, riveting down. But I was six.

What had she said? If I moved, she would have to come back...How would she know? I shifted position just a little bit. Nothing happened. I moved again, then more and more, thrashing around like a skinny whale. Sure enough, the ice began to melt. I heard it splashing onto the floor. I was giggling, having fun.

Then the squeak of rubber-soled shoes, coming fast. The curtains ripped open.

Big red face up close to mine, then *BAM* she slapped me side of the head.

"I <u>told</u> you not to move," she explained.

I never told my parents about it, being reasonably certain they would side with vested authority.

There would be one more encounter with the evil nurse, which will bring us (I promise!) to the point of this story.

A newcomer on a wheeled bed was shoved in next to mine: a pale little boy who never opened his eyes. But what caught my attention was the glass jar of red liquid suspended above him, with a plastic tube going down into his arm.

I was scared of the nurse. But I was eaten up with curiosity.

"What is that?" I finally asked about the red jar, bracing myself for an explosion.

The nurse looked over at the pale kid, then to the jar, and back to me. And she smiled...

"He gave trouble," she said, "So *we're taking his blood out.* You're not going to give trouble, are you?"

In a couple days the pale boy went away. I never knew what happened to him. But I had nightmares about that nurse, and her vampiric blood jar...

You know, of course, what I did not: the pale boy was just having a transfusion. They were putting blood in, not taking it out. (This was 1951, and some hospitals used glass jars instead of blood bags.)

But what if there was no way of storing blood to use in the transfusions?

Several countries worked on the problem. Russian scientists systematically stored blood for wounded soldiers.

"...Vladimir Shamov and Sergei Yudin showed in the early 1930s that blood from cadavers could be briefly preserved, leading to..."canned" blood at refrigerated blood centers in Leningrad."[1]

Their work was studied by a wonderfully kind man, Dr. Bernard Fantus, when he worked at the Cook County Hospital in Chicago, Illinois. He married a nurse (one of the nice ones!) named Emily Senn, and they adopted a little girl, Ruth.

Fantus began thinking about the need for blood, portable and storable, for all those wounded soldiers. He brooded about the possibilities.

He talked about it with his daughter, and one day little Ruth looked up and said:

"Why not call it a blood bank?"

In 1937, Dr. Fantus developed the world's first blood bank. When World War Two broke out, many lives were saved, because of stored and transfused blood.

But what if there was no blood match between donor and recipient? Or if the blood was tainted by disease? Testing could detect some unsafe blood, but maybe not all.

What if, instead of borrowing blood from others, it was made — from stem cells?

Dr. Tannishtha Reya of UC San Diego wanted to take on that challenge — if the funding could be found.

One of the most interesting grants given by the California stem cell program is the "Inception" award. This award "provides seed funding for great ideas that have the potential to impact human stem cell research, but need some initial support. It is hoped this will enable the researchers to test their ideas, and give them the data they need to compete for more substantial funding." — CIRM Stem Cell Agency, May 25, 2017.

Dr. Reya won an "Inception" Award, $223,200, and she focused on developing a stem cell-derived source of red blood cells. The idea seems vital to me.

"Every two *seconds* someone in the U.S. needs a blood transfusion. But sometimes, due to a shortage of donors, there is not enough blood... and surgeries... have to be canceled.

---

[1] https://www.chicagotribune.com/business/blue-sky/chi-blood-bank-bernard-fantus-bsi-series-story.html

"Creating a safe, unlimited supply of universal donor blood cells could have a huge impact, not just in the U.S., but worldwide," said Jonathan Thomas, PhD, J.D., Chair of the CIRM board of directors.

P.S. One more story about the kind Dr. Fantus? He did not want to see children suffering, struggling not to swallow hateful-tasting medicine. Why did they have to suffer? He kept thinking — why must medicine taste bad?

So Dr. Fantus worked with confectioners, professional candy-makers — and came up with flavors more acceptable to kids…

Medical research: organized kindness — a way to ease suffering and save lives.

35 Someone Who Gets Things Done

I have always been fascinated by people who get things done...like Maria Bonneville. (CIRM photo.)

I have always been fascinated by people who get things done, often very quietly, moving behind the scenes, sometimes little-known, except by leadership.

Such a person is Maria Bonneville, Executive Director to the Governing Board. If you come to the meetings of the governing board, the ICOC, you will meet her right away, because she will say "Hello!" to you first. A bright, cheerful, athletic person, she practically bubbles with enthusiasm for her job. Here is a chance to say "Hello!" to her!

Maria, I regard you as the ultimate follow-through person. When you say something is going to happen, it will definitely get done.

How are you so organized?

Thanks! I use lists a lot. Also whiteboards. I scribble a chore on the board, and when it is done, I erase it.

How would you describe what you do?

I started as a conduit from the ICOC board to CIRM and vice-versa. I had to make sure that what the board got what they needed: documents, conversations, new contacts — fresh eyes and ears. Now the job has become bigger, and I do more work on communications: more outreach to advocates. There are also internal service jobs; we create and assist other projects to add value, and make sure we stay mission-oriented.

Our job is to provide therapies for unmet medical needs, to bring relief to the suffering. We want to always seek to accelerate the development of treatment, seeking improvements in efficiency, making sure the research happens carefully at all times, but without needless delay.

Also, it is vital that we learn from the patients, and that means maintaining conversations with them, as the ultimate consumers. CIRM has brought patient advocates and scientists under the same roof, systematically. Before CIRM, scientists might go their whole careers without any actual contact with people who had their point-of-interest disease. We are building a community between patients and scientists. At the ICOC meetings, you will see this in action. The scientists will talk, and so will the patients, as active participants. Anyone who attends is welcome to speak.

What advice would you give to patient advocates?

Right now, there are 1,200 patients enrolled in clinical trials — my advice? Don't stop! Keep fighting for what you need, working as a disease group, cooperating with researchers.

I did not realize the importance of this until my father became sick with Parkinson's. He died of it, as did several other members of my family. What I learned from this was how very difficult it is to obtain even the most basic services. Who should we turn to for guidance? Like how do you get a car with wheelchair accessibility? Today, I have contacts. I know who I would call, but this had to be learned. It is so important to have people around you who know what needs to be done, and how to do it.

I would like to see our Alpha Clinics come up with a model of "patient navigation", sharing awareness. Find out how to get someone assigned to you, to point out what needs to be done next. Like my mom could not keep track of everything, and my Dad was just overwhelmed. So we do need an industry standard.

For instance, suppose you want to be in a clinical trial, and it might take a week — how do you afford that? There are groups who will help with the costs, like the Lazarus Foundation, but you have to know about them. Naturally we cannot take away from the funding of our research, but we could set up a model program.

I love a mission-driven organization like CIRM. It is very clear what we are here to do, how could we not relate to it? This is not just pushing buttons, this is something really worthwhile.

Greatest challenges?

Not losing focus: there are a million things to do; maintaining focus is sometimes difficult. We need to keep thinking about funding good science. Also, sometimes it is hard for the board to say no! We need to move the entire field forward.

CIRM supports the efforts of many women scientists — what advice or encouragement might you give to girls or women considering a career in science?

Find a mentor! Identify someone you respect, and ask if they would consider mentoring you: take you under their wing, so to speak.

What progress in fighting diseases or conditions most excites you?

Gene therapy! We recently started funding more gene therapy, keeping us on the edge of science advancement.

What would be lost, if CIRM was not renewed?

So much! It would be such a shame if California let all this good progress go. A great deal of very promising science is partially begun, but has not reached the point in its growth where industry could fund it.

By funding early stage research, we "de-risk" it, so that the most promising methods will stand out. Without CIRM, that would stop. Of course, not everything funded will succeed. We need to know what will not work as well, to prevent wasting time and money on avenues that will not succeed.

But for science to move forward, we have to put money out there — it would be a huge shame to not see the value develop.

36 The Voice of CIRM

Kevin McCormack is whiskery, cheerful, always ready for a laugh — but absolutely no clown. (CIRM photo.)

When you next come to a meeting of the California stem cell program, there is someone you should definitely meet — and almost certainly will, because he will seek you out, and shake your hand.

Meet KEVIN MCCORMACK: Senior Director of Public Communication and Patient Advocate Outreach, whom I consider the voice of CIRM.

How did you get started in this field?

I was a journalist, covering health and medical issues at KRON TV in San Francisco. In 2004 I covered Proposition 71, and thought: what a cool idea! When I heard there was a job opening, I applied for it, and here I am, 7 years later.

Your cheerful attitude is one of CIRM's great assets — what is your secret?

I tried being miserable once and it just wasn't very much fun. Seriously, I feel lucky to have work like this — seeing history being made, and spending time with amazing researchers, patients and advocates. Also, when you see what some of the patients put up with, none of the rest of us have a right to feel sorry for ourselves — it's a privilege to work with such people.

How much time do you spend on the phone every day?

It depends, but generally about an hour a day. It may be people looking for clinical trials for themselves or family members. I just guide them a little. Then of course there is outreach to patients, reporters and scientists. I need to stay on top of the science, as the field is moving, sometimes bewilderingly fast.

Your writing is exceptionally clear and easy to follow — do you have any writing tips to share — especially scientists, who sometimes use too many big words?

For scientists, it is hard. They work with other scientists in a shared language. For them I might suggest, how would they explain what they do to a 10 year old child? Try to speak in words a person that young could understand. This is vital. In publicly funded research, the public has a right to know what's going on, and that means in language they can understand.

If a scientist wants a grant from CIRM, what should he or she do first?

Above all, come to the website! It is all laid out there, to make it convenient for them. First, they need to see if they are eligible. The applications are right there; what they are hoping to do, what we can fund. Just go to the home page: www.cirm.ca.gov. At the top right hand corner of the page are two words: funding opportunities. That is where they start. It is broken down into categories and groups, criteria for eligibility. For instance, click on discovery. That lays out what the grant will cover — what is fundable, and not. It tells you how to apply, application dates, and that all-important contact email. When you have questions, there is our email, just write your question and click send. Someone from our team will respond. We want the scientists to make the best applications they can. If they make the best application they can, that gives them the best chance to have their work succeed — and become something better than hope.

California does not want our money wasted — how does CIRM insure good performance from scientists?

From the beginning of the application process the CIRM team is heavily involved. We start by reviewing the application to make sure it's eligible, then we check other things such as, do they have a realistic and reasonable budget estimate? Are they asking $5 million for a project that can be done for less? We need to know problems like that right away. Then comes our out-of-state experts, the Grants Working Group (GWG). These are people who live by details! They will assess each project on its scientific merits. They also look at the applicant's background, his or her level of experience, to see if they think they are capable of leading this kind of project. The GWG will make its recommendation, which then goes to the board of directors, the ICOC, for the final decision. Assuming success, partial funding will be awarded, not the whole amount right away. Milestones of progress will be established, and funds given as these goals are reached, step by step. If the milestones are not reached, the funding stops. We don't want to waste good money on bad science. If something isn't working, we stop it — so we can reserve the money for a project with a better chance of success.

Name a couple exceptional scientists, and why?

It is almost impossible to select an exceptional scientist, because the overall quality is so high. But one I must mention is Jan Nolta, stem cell program director of UC Davis. Dr. Nolta is extraordinary not just for her personal scientific achievements, which are remarkable, and the Institute for Regenerative Medicine (IRM) which she made with major donations from CIRM and other sources — she is a mastermind for the relentless fundraising that must be done: but also for the incredible team she has built. She never flags in her efforts, and is never less than caring — incredibly dedicated to the patients; that is Dr. Jan Nolta.

Or Henry Klassen, of the Gavin Herbert Eye Institute at UCI. He could have made a ton of money as a doctor — instead he worked towards the cure for retinitis (blindness), year after year, until now when it is finally paying off. His goals extend beyond the pocket book. He is working to achieve lasting good — to help people who have been told there is no hope.

What do you feel is the greatest impact of CIRM?

We cannot know that yet; it is too soon. But we are continually advancing science. California has become the global center for regenerative research: a pipeline for projects. If a basic or Discovery stage project is successful, then it moves to the Translational stage, if it succeeds there, then it's on to human trials — but the end result must benefit patients. As a group of people, we think about that constantly — everything we do must benefit patients — if it doesn't, all else is meaningless.

This is a time of crisis for CIRM: what would be lost, if California does not renew?

Without renewal, we would lose a chance at bringing promising research to clinical trials, human trials, to see if they can really help people. Deep-pocket investors are needed to ensure a project gets all the way through the clinical trial phase, but they usually hold back investing until the project has results, and shows it might work. They want results from human trials before leaping in with major investments of money. CIRM may provide $5 million for a human trial, to help the project reach the level of success which will justify major investments from others. Without CIRM, it will be extremely difficult for many projects to get to that point. At best, we will lose time, which some patients just do not have.

There are also conditions called "orphan diseases" which affect small numbers of patients and which businesses may not see as big enough to offer a return on their investment. They may not want to risk millions of dollars to do preliminary testing — but CIRM will. Our mission is patients, not money. Remember when Geron pulled out of stem cell research to fight spinal cord injuries — they left the field for purely financial reasons. CIRM was there to help fill the gap with funding, and today, Asterias Biotherapeutics is going forward with that stem cell therapy. Because of that, people left paralyzed by an injury are regaining the use of their arms and hands. They are able to lead a more independent life and are not completely dependent on others for round-the-clock care.

Other points you would like to mention?

The people of California should feel proud of the risk they took. Stem cells were once thought to be fringe science. 2004 was very definitely early days, and we did not know how long it would take to turn the promise of stem cells into reality. But California voters felt the game was worth the gamble, and today, because of their courage and foresight, it is paying off. We all have a stake in this, because we all have a loved one with a chronic illness or injury. Now we can see results, and feel confident that renewing the agency will make a huge difference in people's lives.

37 Adventures on Bridges: Humboldt State University

"After basic training at the college", explained Amy Sprowles", you would receive a grant (roughly $40,000) for a one year internship at a world-renowned stem cell research facility...." (Photo by YouTube.com.)

Imagine yourself as a California college student, hoping to become a stem cell researcher. Like almost all students, you are in need of financial help, and so you asked your college counselor if there were any scholarships available.

To your delight, she said, well, there is this wonderful internship program called Bridges, funded by the California Institution for Regenerative Medicine (CIRM) which funds training in stem cell biology and regenerative medicine — and so, naturally, you applied...

What would this be like? If you were accepted, how might your life change?[1]

After doing some basic training at the college, you would receive a grant (roughly $40,000) for a one-year internship at a world-renowned stem cell research facility. What an incredible leap forward in your career:

---

[1] https://www.cirm.ca.gov/our-funding/research-rfas/bridges

hands-on experience (essentially a first job, great "experience" for the resume) as well an expert education.

Where are the 14 California colleges in this program? Click below:[2]

Let's take a look at just one of these college programs in action: find out what happened to a few of the students who received a Bridges award, crossing the gap between studying stem cell research and actually applying it.

(HSU information is courtesy of Dr. Amy Sprowles, Associate Professor of Biological Sciences and Co-Director of the Bridges program at Humboldt State University (HSU), 279 miles north of San Francisco.)

"The HSU Bridges program," says Dr. Sprowles, "was largely developed by four people: Rollin Richmond, then HSU President, who worked closely with Susan Baxter, Executive Director of the CSU Program for Education and Research in Biotechnology, to secure the CIRM Bridges initiative; HSU Professor of Biological Sciences Jacob Varkey, who pioneered HSU's undergraduate biomedical education program," (and of course Dr. Sprowles herself, at the time a lecturer with a PhD in Biochemistry).

The program has two parts: a beginning course in stem cell research, and a twelve-month internship in a premiere stem cell research laboratory. For HSU, these are at Stanford University, UC Davis, UCSF, or the Scripps Research Institute.

Like all CIRM Bridges programs, the HSU stem cell program is individually designed to suit the needs of its particular community.

Each of the 15 CIRM Bridges Programs may fund up to ten paid internships, but the curriculum and specific activities of each are designed by their campus directors. The HSU program prepares Bridges candidates by requiring participation in a semester-long lecture and stem cell biology laboratory course before selection for the program: a course designed and taught by Sprowles since its inception.

She states: "The HSU pre-internship course ensures our students are trained in fundamental scientific concepts, laboratory skills and professional behaviors before entering their host laboratory. We find this necessary since, unlike the other Bridges campuses, we are 300+ miles away from the internship sites and are unable to fully support this kind of training during the experience. It also provides additional insights about

---

[2] https://www.cirm.ca.gov/our-funding/funded-institutions

the work ethic and mentoring needs of the individuals we select that are helpful in placing and supporting our program participants."

How is it working?

Ten years after it began, 76 HSU students have completed the CIRM Bridges program at HSU. Of those, the overwhelming majority (over 85%) are committed to careers in regenerative medicine: either working in the field already, or continuing their education toward that goal.

But what happened to their lives? Take a brief look at the ongoing careers of a "Magnificent Seven" HSU Bridges scientists[3]:

CARSTEN CHARLESWORTH: "Spurred by the opportunity to complete a paid internship at a world class research institution in Stem Cell Biology, I applied to the Humboldt CIRM Bridges program, and was lucky enough to be accepted. With a keen interest in the developing field of genome editing and the recent advent of the CRISPR-Cas9 system, I chose to intern in the lab of a pioneer in the genome editing field, Dr. Matthew Porteus at Stanford, who focuses in genome editing hematopoietic stem cells to treat diseases such as sickle cell disease. In August of 2018 I began a PhD in Stanford's Stem Cell and Regenerative Medicine program, where I am currently a second-year graduate student in the lab of Dr. Hiro Nakauchi, working on the development of human organs...The success that I've had and my acceptance into Stanford's world class PhD program are a direct result of the opportunity that the CIRM Bridges internship provided me and the excellent training and instruction that I received from the Humboldt State Biology Program."

ELISEBETH TORRETTI: "While looking for opportunities at HSU, I stumbled upon the CIRM Bridges program. It was perfect — a paid internship at high profile labs where I could expand my research skills for an entire year...the best fit (was) Jeanne Loring's Lab at the Scripps Research Institute in La Jolla, CA. Dr. Loring is one of the premiere stem cell researchers in the world... (The lab's) main focus is to develop a cure for Parkinson's disease. (They) take skin cells known as fibroblasts and revert them into stem cells. These cells, called induced pluripotent stem cells (iPSCs) can then be differentiated into dopaminergic neurons and transplanted into the patient.... My project focused on a different disease: adenylate-cyclase 5 (ADCY5)–related dyskinesia. During my time at Dr. Loring's lab I learned incredibly valuable research skills. I am now working in a mid-sized biotech company focusing on cancer

---

[3] http://www2.humboldt.edu/biosci/programs/grad.html

research. I don't think that would have been possible in a competitive area like San Diego without my experience gained through the CIRM Bridges program."

BRENDAN KELLY: "After completing my CIRM internship in Dr. Marius Wernig's lab (in Stanford), I began working at a startup company called I Peace. I helped launch this company with Dr. Koji Tanabe, whom I met while working in my host lab. I am now at Cardiff University in Wales working on my PhD. My research involves using patient-derived neurons (iPSC) to model Huntington's disease. All this derived from my opportunity to partake in the CIRM-Bridges program, which opened doors for me."

SAMANTHA SHELTON: "CIRM Bridges provided invaluable hands-on training in cell culture and stem cell techniques that have shaped my future in science. My CIRM internship in John Rubenstein's Lab of Neural Development taught me amazing laboratory techniques such as stem cell transplantation as well as what goes into creating a harmonious and productive laboratory environment. My internship projects led to my first co-first author publication.

After my Bridges internship, I joined the Graduate Program for Neuroscience at Boston University. My PhD work aims to discover types of stem cells in the brain and how the structure of the brain develops early in life. During this time, I have focused on changes in brain development after Zika virus infection to better understand how microcephaly (small skulls and brains, often a symptom of Zika — DR) is caused. There is no doubt that CIRM not only made me a more competitive candidate for a doctoral degree but also provided me with tools to progress towards my ultimate goal of understanding and treating neurological diseases with stem cell technologies."

DU CHENG: "Both my academic and business tracks started in the CIRM-funded...fellowship (at Stanford) where I invented the technology (the LabCam Microscope adapter) that I formed my company on (iDU Optics LLC). The instructor of the class, Dr. Amy Sprowles, encouraged me to carry on the idea. Later, I was able to get in the MD-PhD program at Weill Cornell Medical College because of the invaluable research experiences CIRM's research program provided me. CIRM initiated the momentum to get me where I am today. Looking back, the CIRM Bridges Program is an instrumental jump-starter on my early career...I would not remotely be where I am without it..."

CODY KIME: "Securing a CIRM grant helped me to take a position in the Nobel Prize winning Shinya Yamanaka Lab at the Gladstone Institutes, one of the most competitive labs in the new field of cell reprogramming. I then explored my own reprogramming interests, moving to the Kyoto University of Medicine, Doctor of Medical Sciences Program in Japan, and building a reprogramming team in the Masayo Takahashi Lab at RIKEN. My studies explore inducing cells to their highest total potential using less intrusive means and hacking the cell program. My systems are designed to inform my hypotheses toward a true お好みの細胞 (okonomi no cybo) technology, meaning 'cells as you wish' in Japanese, that could rapidly change any cell into another desired cell type or tissue."

SARA MILLS: "The CIRM Bridges program was the key early influencer which aided in my hiring in my first industry position at ViaCyte, Inc. At that strongly CIRM-funded institution, I was ultimately responsible for the process development of the VC-01™ fill, finish processes and cGMP documentation development. Most recently, with over two years at the boutique consulting firm of Dark Horse Consulting, Inc., I have been focusing on aseptic and cGMP manufacturing process development, risk analysis, CMC and regulatory filings, facility design and project management to advise growing cell and gene therapy companies, worldwide."

Like warriors fighting to save lives, these young scientists are engaged in an effort to study and defeat chronic disease. It is to be hoped the California stem cell program will have its funding renewed, so the "Bridges" program can continue.

For more information on the Bridges program, which might help a young scientist (perhaps you) visit the following URL:[4]

One closing paragraph perhaps best sums up the Bridges experience:

"During my CIRM Bridges training in Stanford University, I was fortunate to work with Dr. Jill Helms, who so patiently mentored me on research design and execution. I ended up publishing 7 papers with her during the two-year CIRM internship and helped making significant progress of turning a Stem Cell factor into applicable therapeutic form, that is currently in preparation for clinical trial by a biotech company in Silicon Valley. I also learned from her how to write grants and publications, but more importantly, (to) never limit your potential by what you already know." — Du Cheng

---

[4] https://www.cirm.ca.gov/our-impact/internship-programs

38 Inside Gloria's Heart

Gloria, beloved wife of 50 years ... at risk of heart failure? (CIRM photo.)

In her hospital gown, Gloria, my wife of half a century, was walking on the treadmill.

Gradually the speed of that walking belt increased, faster and faster. Worried she might slip, I crowded in closer to her.

With a hum, the machine altered its angle — forcing Gloria to trudge up a hill.

Wires were taped to her chest. These led to a machine, behind which sat a cardiac specialist, studying the messages from Gloria's heart.

She breathed more rapidly. Sweat glistened.

The doctor frowned.

"I don't like this," she said, looking at the numbers on the treadmill screen.

At 74, Gloria was a prime candidate for a heart attack. Her weight was a challenge; and her relatives died frequently of cardiac arrest.

Stents were mentioned: multi-purpose tubes down the arteries to her heart. A tiny camera would send pictures. If it showed a narrowing of the artery, a metal mesh stent would be inserted, permanently, more room for blood-flow.

"If there are major problems," said the doctor, "You might need to consider open heart surgery."

"None of that!" snapped Gloria. Her mother Soledad had had the procedure: her ribs cracked open, exposing the heart. Afterwards the ribs were sewn back together and Sally lived to 86, so it was a victory.

But the healing, Gloria remembered, had been a year-long agony.

"No open-heart surgery!" she repeated, looking straight at me.

She did give me the power of decision (about the stents) for when she was unconscious, but her wishes about open-heart surgery were clear.

"Not even if it would save your life?" I asked.

"Not even then."

Two days later we arrived at the hospital: eight in the morning.

Would we be first in line for the operation?

"Code blue!" came the voice on the loudspeaker.

That meant someone was having a heart attack, right now.

He or she got the first operation. That was right and proper, (being at risk of an immediate death) but still it increased the nerve-wracking delay. "Code blue" happened again and again.

Gloria is not over-burdened with patience. We often say, when they were passing out patience, she had no time to wait for any.

I asked the nurse how long the wait would be.

"No way of telling," she said, "This is a hospital, you know. Dealing with emergencies is what we do all day long."

"What if the doctors get tired?" said Gloria. "Have they had lunch? I don't want someone exhausted working on me! I want them well-fed and rested!" The nurse reassured her; there were five doctors on duty: plenty for the job.

Nourishment was on Gloria's mind, having fasted (neither food nor water) since midnight. Now, they would not even give her ice chips. Thirst became a burning desire, made more as she is a diabetic.

"I need to go to the bathroom!" she said suddenly. This was strange, because she had just been.

"Of course — and I will go with you," said the nurse. Gloria tried to talk her out of it, saying she had been using the restroom on her own for quite a while now.

"No, no, it's okay," said the nurse, and accompanied my wife into the restroom.

Gloria came back looking frustrated.

"I was going to cup my hands at the faucet and get some water to drink!" she whispered, "I think she was on to me!"

"They do this so often, they must know all the tricks!" I whispered back.

I read out loud from a Maigret mystery book by Simenon, a wonderful author. But it was over too fast. His books are short, and I only had a few chapters to read.

And whenever I stopped talking, I felt the fear closing in.

A phrase from Edgar Allan Poe returned to me, "Darkness, decay, and the red death held illimitable dominion over all."

As far as I could tell, Gloria was not visibly concerned about dying.

"What better place than to be with God?" she said when we discussed it.

"Well, that's nice, but what about me?" I asked.

"We would only be separated for a little while", she said, "You are getting pretty old, after all."

That comforted her, but not me.

She told me to put my head down on the bedrail beside her, and take a nap.

I tried, but the bar was cold metal, and nightmare scenarios played in my mind.

What would I do if I lost her? We were best friends for half a century, and we still played cards every night. She hates to lose and seldom does — but in our last game I kept getting great cards and there was no way I could lose. I deliberately made stupid moves one after another, but it was almost impossible not to win. I managed to lose somehow, and she laughed her laugh of sheer delight.

I knew I could physically survive without her, eating pizza and cold cereal.

But it would be like going from movie Technicolor to black and white...

Once, I stopped an airplane for her. She had been visiting my sister Patty. I was home on leave from the Army. Gloria was with another male. But when I saw her, I physically stepped between her and the boyfriend, turning my back on him. Somehow Gloria and I ended up in the same car heading for the airport, while boyfriend rode with my mother. I had always been shy before Gloria, but now I was talking so much we missed the freeway turnoff, and arrived at the airport — late.

The boyfriend put out his arms. "Oh, I was so worried!" he said, but I just wanted to get Gloria on that plane.

I saw a jet — her flight — moving slowly down the runway.

I leaped over a velvet rope, pushed a door open, and ran out on the tarmac, fast as I could, running beside the accelerating airplane, shouting: "Stop the plane!"

A tiny face appeared at a window. I waved, frantically. The face went away. I thought I had lost but kept running. The plane separated the distance between us — I could not catch up — and then its brakes squealed, and the airplane stopped, and Gloria came out, and got aboard. Today I would probably have been shot by an air marshal, but those were gentler times.

If Gloria died? I imagined myself wandering through our empty house, searching for her, calling "Hon?" but there was no answer.

I raised my head, looked at her: silent, immobile, eyes shut: like in a coffin.

And then, just for a second, and I don't know why — my selfish fears went away, and I thought: hundreds of millions of people all across the world; all loving someone like I love her. How many might needlessly die from heart disease?

A scientist's name flashed through my mind. Joseph Wu. I had written about him several times, how he (from Stanford) and Deepak Srivastava (Gladstone Institutes) were fighting end-stage heart failure by "replacing lost cardiomyocytes (heart cells) with healthy ones...."[1]

---

[1] https://www.cirm.ca.gov/our-progress/awards/human-embryonic-stem-cell-derived-cardiomyocytes-patients-end-stage-heart-0

CIRM (the California Institute for Regenerative Medicine) was providing grant money for the two great scientists, money they needed to fight heart disease.

But when that program's money was gone...would California renew it?

We must make it happen.

The curtains yanked back. "Your turn!" said the smiling male nurse. He put his hands on the end of Gloria's wheeled bed.

"Wait!" I said, "This is the most wonderful person in the world — be careful!"

"I will, I promise!" he said — like what else could he say? — and wheeled her away.

My last words (last words!) to Gloria were: "Love you, angel!"

Hers were more practical: "Go get something to eat!" like I might not remember.

After she left I found the cafeteria, ate a miserable slice of alleged pizza. It was semi-warm, with little bumps and it looked like a model for some rare disease.

I went back to the waiting room, and stayed there for an eternity or two.

The door opened. It was Gloria's surgeon, still in his blue clothes.

He had a serious expression on his face.

I have done construction work on a high-rise building, when it was just the metal skeleton. Sometimes two beams did not quite fit together, so one was slightly lower than the next, and you might step backwards and "fall" six inches — a very short fall but packed with terror — that's what I felt when the surgeon came in.

He was going to tell me they lost her.

But he said, "No, no, everything is okay!"

"What, what, tell me!" I said, not quite grabbing his shirt front.

"I do not often get to give such good news," he said, "But Gloria's veins are fine — everybody on this floor would love to trade veins with her!"

I shook his hand several times, after making him repeat himself. There was one vein that was 20% narrower than it should be, (doubtless the reason for the original concern) but it was not life-threatening.

No need for stents, nor surgery.

My partner in life was not going anywhere...

"I'm thirsty," said Gloria, when she woke up.

She inhaled three bottles of water, two cranberry juice boxes, some vile "tropic-flavor" gelatin, then a lunch, or maybe two: everything she asked for.

She was on the phone with our children immediately, and while they were talking, the words "Code Blue" entered the room again — but not for her.

Not for my Gloria.

39 Fighting Beside Other Countries

Tony Clement, Minister of Health. Canada brings $100 million to the CIRM/Canada fight against cancer.... (Wikipedia photo.)

I love international cooperation. War, generally, seems like the very definition of failure. But when countries can work together?

First, though, I must be clear about CIRM. With rare exceptions, every dollar of the California stem cell research program is spent <u>in California</u>. The exceptions are for equipment or materials that cannot be had for reasonable prices in California. Other than that, our home state money is spent at home.

But what if one of our scientists wanted to do a team project with an expert in another country?

There are established pathways for this.

First, the out-of-state scientist can move here. He/she can set up headquarters and apply for CIRM funding, take his/her chances like everybody else.

Or, an agreement can be drawn up between California and the other state or nation. Our scientist applies for funding here; the other scientist does the same from his or her place of residence. If the project and funding are approved, they go ahead.

Great minds combine — and we get double bang for the research buck!

This can be other states, nations, or research organizations. There are already more than a dozen established cooperative agreements in place.

Bob Klein structured the California stem cell program to allow for international cooperation.

Which partnership has the most potential? I caught up to Bob, the man who began Proposition 71 (the initiative which led to the California stem cell program) and asked his opinion on *which country was set up to do the most good work*, in coordination with the California stem cell program:

"Canada," he said immediately, "Canada."

I looked it up, and this is some of what I found, a treasure trove of information:

On June 18, 2008, Tony Clement, Canada's Minister of Health announced:

"Canada will contribute more than $100 million to the Cancer Stem Cell Consortium, (to) work with the California Institute for Regenerative Medicine (CIRM)..."[1]

To raise one million dollars is a triumph — but one hundred million? That money would not have been there, without the existence and backup of California's stem cell program. Our neighbor to the North is a leader in stem cell and cancer research, but the partnership made dedicating big money easier.

Then-Governor Arnold Schwarzenegger said it well:

"California is committed to being a leader in stem cell research, but no one state or nation should do this alone. Collaborations such as this,

---

[1] https://www.cirm.ca.gov/about-cirm/newsroom/press-releases/06182008/minister-clement-governor-schwarzenegger-join-forces

which bring together leading medical research capabilities, have great potential in improving the lives of not only Californians, but people around the world."[2]

More examples? Here are my non-scientist interpretations of a few cooperative projects our various countries/states/organizations are doing. As always, when I say "we" or "our," I mean the California stem cell program, not me personally; I have no connection with it, except pride.

California pays for California scientists; the other countries pay for theirs. Here are some of our cooperative projects.

Working with France, we are trying to understand how the body's stem cells become what kind (differentiate) and how many are made.

With Victoria, Australia — working to ensure safety of cell therapy.

Andalusia, Spain — trying for a treatment of critical limb ischemia (to lessen potentially fatal obstruction of arteries, and needless amputation of limbs).

UK — Targeting leukemia stem cells, the disease which killed so many, including my sister, Patricia C. Reed.

NIH — developing a drug testing panel for autism...using stem cells to find out what affects that condition.

JDRF — working side by side to develop a stem cell therapy for diabetes.

Germany — Developing a stem cell liver support system.

Australia — Stem cell lines to make blood cells, to replace blood lost in operations.

Japan — studying the micro-environment in stem cells, to help fight cancer.

Maryland — developing human nerve stem cells to treat brain injury.

Canada — Targeting the actual tumor-making cells, within the cancerous tumor.

China — developing cells to make a liver-substitute.

Every state and nation has its strengths: unique and irreplaceable. Together, we are so much more.

Should we not work for the health of every person everywhere, no matter the accidental location of their birth?

---

[2] https://www.cirm.ca.gov/our-funding/collaborative-funding-partners

40 Jobs and New Money

Kathy Ivey is Research Director at Tenaya Therapeutics — her story connects with $3.14 billion (extra) for California. (Tenaya Therapeutics photo.)

Imagine a government program that *attracts money*, instead of only spending it...

The California Institute for Regenerative Medicine (CIRM) has spent approximately $2.6 billion since its beginning.

But from other sources, **the program has attracted an additional $3.14 billion** — and more is almost certainly on the way.[1]

Did anyone vote for the California stem cell program because we thought it might be a "cash cow"? Probably not. The job of the program boils down to six words: to ease suffering and save lives. For me, that's everything. That's why I write; it's pretty much why I am alive. Millions

---

[1] https://www.cirm.ca.gov/sites/default/files/files/about_cirm/CIRM%20 2018%20Annual%20Report.pdf

of people suffer from chronic disease and disability. That includes my paralyzed son Roman Reed, who goes through hell every day. If you see him on the street, you would not know. He smiles and keeps his troubles to himself. But I am his father and cannot look away.

The California stem cell program is fighting for cures: systematically attacking medical enemies: illness and injuries long considered incurable. Right now there are more than 40 clinical trials, testing stem cell treatments against disease and disability. That is the purpose of CIRM.

But there is also a financial return.

No, not the almost $200,000 royalty check CIRM just received from City of Hope.

"The royalty check, $194,345.87, was from royalties generated from a $5.2 million award in 2012 to the City of Hope for research involving a potential therapy for glioblastoma, one of the deadliest forms of brain cancer." — Kevin McCormack, CIRM Director of Communications.

But there is a far more significant financial benefit, and it is happening right now.

CIRM helps the economy, by attracting new money.

Here's how:

1. CIRM provides initial funding for researchers with new ideas: people like Kathy Ivey, who moved from Texas to work with Deepak Srivastava at Gladstone Institute in California. (For the story of her training grants, see my 2015 book, STEM CELL BATTLES.)
2. If the scientists' ideas prove successful, CIRM may help pay for the long series of FDA tests: as it did for Gladstone's heart regeneration efforts.
3. The private sector may get involved. When companies are formed, theories may become therapies, available to the public. One such company called Tenaya Therapeutics was just formed to develop heart repair therapies developed with the help of CIRM, in this case from within Gladstone — with the leadership of Srivastava and Ivey.

Here is Dr. Srivastava: (as he speaks, listen for the $50 million dollars).

"CIRM has funded the full pipeline of our work on cardiac (heart) regeneration — from basic discoveries, all the way to preclinical studies. As a result of their support, we established Tenaya Therapeutics, a local startup company that launched with $50 million in Series A investment and aims to tackle heart failure." — Deepak Srivastava, M.D., President, Gladstone Institute.[2]

That $50 million dollars is new money: California jobs. CIRM started something, and then private enterprise added to it.

They call that leverage; the encouragement of new business — like biomedicine, the science of life. Did you know that life science has become the second largest industry in California? That it is ahead of aerospace, movies and the internet? Only the computer industry is larger than the industry of cure.[3]

This means jobs: not minimum-wage starvation stuff, but good-paying, valuable, life-affirming jobs. Not just scientists and lab technicians, and the people who administer the program, but construction, raising new structures where nothing was before: the twelve new stem cell research sites, up and down the state — and the business people who provided their supplies.

When the nation was in recession, CIRM provided a jolt of new jobs, and it continues to do so today.

Let's be exact: here are statistics from the CIRM annual report, 2018–2019.

"Co-funding: **$1 billion**: Funding from institutions, industry or investors who join with CIRM to fund a specific project at the outset."

"Partnership Funding: **$1.6 billion**: Disclosed support committed by partners independent of CIRM funding to help advance a project."

"Additional Funding: **$541 million**: Any funding a Principal Investigator can secure by leveraging CIRM's backing of the project to attract additional funds from investors." — page 6 of the annual report.

There it is, **$3.14 billion** in new money, which came to help CIRM's effort: that is benefiting California right now in jobs and revenue — in addition to the fight for cures.

P.S. Pride compels me to mention that a paragraph of mine was used in CIRM's annual report, on page 24, inspiring the title: "Something Better than Hope."

---

[2] https://gladstone.org/about-us/news/tenaya-therapeutics-launches-goal-curing-heart-disease

[3] https://cdn2.hubspot.net/hubfs/2379287/CLSA-&-PwC-2018-CA-Life-Sciences-Industry%20Report-Final.pdf?t=1511309908554

"Today, thanks to the 7.2 million voters who authorized the California Institute for Regenerative Medicine, or CIRM, we have **something better than hope**; we have results, accomplishments, people made well — and a systematic way to fight chronic disease." — Don C. Reed, vice President, Americans for Cures Foundation.

41 Scars: By Moray Eels and Other Causes

What is the connection between a moray eel bite — and Maksim Plikus' fight for scar-less wound-healing? (UCI photo.)

For a writer, scars mean stories. For instance, on the ring finger of my right hand, there is a tiny white scar, from when I was a diver at Marine World long ago.

I had been scrubbing algae off the walls in the moray eel tank, inhabited by several dozen of the snake-like fishes. Breathing with their mouths open, they showed teeth like inward-pointing slivers of glass. The water grew murky, as I scoured around the little caves full of eels. Visibility faded. I could see the sides of my nose and the lining of the plastic mask and then even that went away.[1]

I felt a tension rising in the water, like the music in the horror movies when something bad was about to happen. And then it did: like a mouse

---

[1] http://tinyurl.com/y3399pxg

trap snapping on my finger; I was yanking my hand up and down, trying to get free.

It is said that when an eel bites down, its jaws lock. Fortunately, this is not true.

When the eel let loose. I climbed out of the tank, cleaned the wound, went to the doctor. He did not believe me, until he saw the X-ray.

"It is a classic avulsion fracture!" he said, proud as if he had done it himself, "It penetrated the bone!"

I could have told him that, having cleaned out the bite with a broom straw, which went all the way through the finger and out the other side…

Scars are living zippers to close wounds.

They can also be harmful. In spinal cord injury, a scar can block information sent by nerves from brain to body. When my paralyzed son Roman Reed tells his leg to move, the signals do not get through — the leg does not move.

A scar on the liver interferes with its function.

Multiple sclerosis, which means literally "many scars," can affect brain and spine.

And too much scarring on the heart can end a life.

But what if there was a way to heal torn flesh *without the scar* — or turn an existing scar into smooth seamless skin?

On opposite sides of America, scientists are working to reverse the process of scarring: Drs. Maksim Plikus, a scientist at the University of California at Irvine, and George Cotsarelis, of the University of Pennsylvania. Dr. Plikus received a CIRM training grant as a postdoc at USC.

It seems there are important differences between normal skin and a scar.

What is the recipe for normal skin? Primarily: hair follicles, oil- and sweat-making glands, collagen, and numerous fat cells.

And a scar? Just use the collagen, leave out the hair, glands, and fat.

What if those missing ingredients could be inserted into a wound?

According to the scientists' joint paper in *Science*, when those conditions are met, "Adult mice can…regenerate nearly normal-looking skin."

Remember that great line from the movie FIELD OF DREAMS? "If you build it, they will come"? Well…

"If you build the hair, the fat will follow."

Hair follicles produce growth factors. These make the fat needed for scar-less wound healing. In their most recent 2019 paper in *Nature Communications*, the authors showed that new fat cells can regenerate.

In science-talk:[2]

"Hair follicles activated (a) protein (BMP) signaling…(fat cell)…factors in the myofibroblast (scar-forming) cell. Thus, it may be possible to *reduce scar formation* by adding BMP." (emphasis added — DR)

The potential benefits are enormous. In addition to its hoped-for effects on heart attacks, liver malfunction, and spinal cord injury — anti-scarring therapies may also help reverse lung fibrosis, even fight off infections.

The Plikus research has progressed, and may now be close to human application, and eligible for CIRM funding.

Unfortunately, CIRM is running out of money. Begun in 2004 (actually 2007, after all the conservative lawsuits had been defeated) CIRM was designed to last ten years. Thanks to careful spending, the funding has been stretched out well past its planned end point, but still the funds are running low.

Will there be a renewal of the program?

That is my hope: that when Dr. Plikus is ready to apply for his next major grant, CIRM will be there to answer his request.

---

[2] https://www.ncbi.nlm.nih.gov/pmc/articles/PMC5464786/

42 Battling Duchenne

"I borrowed $100,000 and set out to find a cure for Duchenne Muscular Dystrophy" — Pat Furlong. (Goldlab Foundation photo.)

Caleb Sizemore approached the speaker's platform, to address the ICOC, the Board of Directors of CIRM.

Two things stood out: an astonishing smile, lit by blazing blue-green eyes — and the carefulness of his walk. He seemed to be setting each footstep deliberately, as if the floor was shifting.

"I am a 21-year-old patient with Duchenne Muscular Dystrophy," he said. "I am involved in the Capricor clinical trial called HOPE, funded and supported by the California Institute for Regenerative Medicine.

Caleb Sizemore meets CIRM's new President Maria Millan and other advocates, including author on right, Adrienne Shapiro on far left. (CIRM photo.)

"Duchenne is a progressive muscle-deteriorating disease. It attacks the legs first, and most people stop walking at 12 years old, so for me it is a miracle to be walking at all. I know there is a wheelchair in my future, pretty soon," he said. "But they have some pretty cool power wheelchairs now.

"After the legs, Duchenne affects other muscles, and eventually scars the heart and lungs, which is what kills you."

He talked about death, and how his religious faith helped him deal with the possibility that he might die as early as 30.

His speech was very matter of fact, cheerful, like discussing a stamp collection.

I have heard many such speeches over the years, from folks with various chronic diseases and disabilities, and their suffering and courage always shakes me.

But I was also waiting for something: hoping he would mention it — the moment he became an advocate.

One of my personal heroines is Pat Furlong, who lost two sons to Duchenne. She fought for them as hard as she could, and when they were gone — she kept on fighting. The moment when she became an advocate? Probably this:

"I borrowed $100,000 and set out to find a cure for Duchenne," she said.

Years of effort passed, and she got the Federal government involved — her work led to an estimated $160 million research dollars for Duchenne. Today Ms. Furlong is an international authority on the disease.

Not everyone can have such incredible success, but to quote her group, Parent Project for Duchenne Muscular Dystrophy: "Nobody can do everything, but everybody can do something."[1]

And Caleb's moment? It sprang from a rough childhood, growing up in South Carolina, teased and bullied because he could not run fast or play sports.

He would not complain about his condition, even talk about it, not wanting to stand out any more than he had to. But when he became a teenager, something happened.

Helping to organize a fundraiser for research, Caleb made a speech before the whole school, and told them about his condition. He had no way of knowing how they would react. But still he spoke, and when he was through...

"They 100% supported me!" he said, and the fund-raiser collected $18,000.

Today that battle continues, a life–death struggle, and Caleb is part of it.

With the help of CIRM, an investigative cell therapy (CAP-1002) has been developed which can mitigate inflammation and fibrosis, potentially slowing the progression of the disease.

Capricor Therapeutics, Inc. is a clinical-stage biotechnology company. Its goal is to commercialize biological treatments for rare disorders.

Capricor developed CAP-1002 to regulate the immune system, reduce scarring and potentially allow healthy new muscle to form.

Caleb received a single dose of CAP-1002. He was part of a 25-patient clinical trial. Thirteen received the medication; twelve received standard care for Duchenne, but no CAP-1002. After 12 months, sustained improvements were reported in the cardiac structure and function of those treated with CAP-1002, compared to those who didn't receive the therapy.

Capricor has now launched HOPE-2, a larger study (up to 84 patients), to further study the safety and efficacy of CAP-1002.[2]

---

[1] https://www.parentprojectmd.org/
[2] http://hope2trial.com/

In scientific terms, it is a "randomized double-blind study" that Capricor says could potentially be a registration trial to bring the therapy to market sooner.

Linda Marbán, PhD., Capricor CEO, noted that the FDA has granted CAP-1002 special designations — the Regenerative Medicine Advanced Therapy (RMAT) classification, and an "orphan disease" classification, both of which enable Capricor to work more closely with the FDA, to help CAP-1002 become available sooner: to patients with Duchenne Muscular Dystrophy.

But would it work for humans? Would it be safe?

Despite preparation and planning, whoever goes first is at risk. Caleb Sizemore was testing the new procedure with his life.

Fortunately, the only negative noted was a temporary irregularity of the heartbeat on several of the patients, at the time of the infusion of the cells, and this did no lasting harm. Marbán has said that Capricor no longer uses the procedure that may have led to the irregular heartbeat.

Six months later, MRI scans revealed that the scarring on Caleb's heart not only stopped increasing, but had grown less.

Caleb knew.

"I could feel my heart beating stronger," he said.

The people on CIRM's board of directors burst into applause.

Caleb Sizemore is a living reason why voters approved the California Institute for Regenerative Medicine, and why it must be supported, protected, and renewed.

Cirm-Funded Projects Affecting Muscular Dystrophy

1. April Pyle, UCLA — $2,184,000
2. April Pyle, UCLA — $2,150,400
3. April Pyle, UCLA — $230,400
4. Alessandra Sacco Stanford (DMD) — $265,500
5. Stanley Nelson, UCLA (DMD) — $86,414
6. Stanley Nelson, UCLA (DMD) — $6,000,000
7. Michelle Calos, Stanford — $1,876,253
8. Michelle Calos, Stanford — $2,325,933
9. Jason Pommerantz, UCSF — $1,381,296
10. Irina Conboy, UC Berkeley — $2,246,020
11. Tiziano Barberi, City of Hope — $1,623,064
12. Judith Shizuru, Stanford — $1,403,557

13. Shyni Varghesi, UCSD — $2,300,569
14. Zack Jerome, UCLA — $1,382,400
15. Julie Baker, Stanford — $2,628,635
16. Kyoko Yokomori, UC Irvine — $632,500
17. Deborah Ascheim, Capricor (DMD) — $3,376,259

Total Expenditures on Muscular Dystrophy: $32,093,200
Total on DMD: $9,728,173

43 In Which Stem Cell Research Saves My Car

Before it was stolen, my car looked like this.... (Photo by Amazon.com.)

One morning not long ago, I was working out at our local gym, lifting embarrassingly light dumbbells, doing sit-ups, pulldowns and other excitement, when someone went into the locker room — and then came out, walking fast.

I remembered him only as someone I had talked to about stem cell research, which did not exactly narrow the field. The California stem cell program is my favorite subject, and if you and I are in the same room for more than a heartbeat or two, the subject will probably find its way into the conversation.

Gloria remembered the stranger's haircut, height, the folded/unfolded status of the clothes he was carrying, and other details.

This would not have mattered, ordinarily.

But when I came out of the shower, my clothes were still in my unlocked locker. So was my gym bag.

But my keys were gone.

Also, my car.

Gloria told me quick, run home, don't let the robber steal anything else!

This did not seem a good idea to me, and I objected. However, being more scared of Gloria than any burglar, I ran the six miles home — well, half I ran, the rest I shuffled. No burglar there, fortunately.

I called the bank to tell them to cancel my credit card, but he (presumably he) had run up $600.

The police told us if the car was not found in 48 hours, we should consider it gone, chopped up for parts (Hondas apparently are valuable for such) and nothing could be done.

At two o'clock in the morning, the doorbell rang.

Two policemen stood outside in the dark.

They had the car and the man who apparently stole it.

Officer Mike C. had driven by a gas station, where he saw a man kneeling at the back of my car. The individual had a Phillips-head screwdriver, and two license plates. When he saw the policeman, the suspect got up and walked rapidly into the gas station's food store, then into the bathroom, where he and my car keys were later found.

The suspect was taken into custody, and identified. Turned out he was on parole for a similar offense. Apparently, he was making a career of car theft, but was just not very good at it.

He pled guilty and had his parole time extended. I did not attend his trial, thinking it unlikely we would become friends, and besides I had already told him about stem cells.

My license plate is STMCLL1 — stem cell one.

The unique plate had attracted the officer's attention.

And that, dear reader, is how stem cell research saved my car.

44 Should Scientists Run for Office?

Shaugnessy Naughton helps scientists — to run for government office. (Daily Kos photo.)

Shaugnessy Naughton, former chemist and breast cancer researcher, ran twice for Congress, to represent the 28th district of Pennsylvania. Sadly, she did not win.

But she gained a perspective which might change politics forever.

If scientists are to be part of the American decision-making process, they should consider getting elected.

Working with political experts like Joel Trippi and Ted Bordelon, Naughton founded a non-profit group (501 c-4) called 314 Action, dedicated to training scientists to run for office.

Where does their name come from? <u>**314 Action**</u> is both an outreach to scientists and an insider's joke.

3.14 is the mysterious formula of pi.

Check this out:

If you take any circle (any one at all) and divide its circumference (distance around) by the diameter (distance across the middle) — *you will always get the same answer* — 3.14.

Try it on a ten cent piece.

The dime's measurements are: circumference, 56.24 millimeters; diameter, 17.9 millimeters. Divide 56.24 by 17.9 and you get... 3.14.

Pi, it seems, is all around us — like science itself.

Why does that matter? Because science is built on provable facts, testable and repeatable evidence — not someone's opinion.

We need people in power who understand that, and live by it.

For instance, on the subject of global warming, it is <u>opinion</u>, not fact, when Lamar Smith (R) says, "Earth is moving out of its heating cycle." Mr. Smith is the Trump-appointed Chair of the House Science, Space, and Technology Committee. He does not believe in global warming. His opinion may be affected, however, by the fact that he has received nearly three-quarters of a million dollars ($727,647) in campaign contributions from the oil and gas industry.[1]

It is a <u>fact</u>, however, when we say, on the 10th of August, 2017, both Kuwait and Iraq recorded a temperature of 129 degrees. Except for Death Valley, California, that is the hottest temperature in recorded history.[2]

And it is a fact that 2014 was the hottest year in recorded history,[3] followed by another world record in 2015,[4] and 2016? Also the hottest year for the planet.[5]

---

[1] https://www.opensecrets.org/members-of-congress/industries?cid=N0000181
1&cycle=CAREER&type=I

[2] http://mashable.com/2016/07/22/middle-east-heat-record/#xno_9Jb6Vmqw

[3] http://time.com/3656646/2014-hottest-year/

[4] https://www.nytimes.com/2016/01/21/science/earth/2015-hottest-year-global-warming.html?_r=0

[5] http://www.politico.com/story/2017/08/10/2016-broke-global-temperature-records-again-241486

Rep. Smith is arguing *against* an enormous report on global warming, and its connection to human activities like pollution. The report is built on "thousands of studies (by) tens of thousands of scientists."[6]

Lamar Smith? "like Trump, Smith rejects the scientific consensus surrounding global warming. He wants to slash federal funding for science agencies."[7]

Does this matter? Only if you believe that devastating weather events like droughts and flooding are affected by global warming — and that pollution caused by oil and coal use has an impact on global warming — not to mention breathing poison is not helpful to your health!

But perhaps there are already plenty of scientists in office, to balance the anti-science politicians?

What professions dominate the House of Representatives and the Senate?

According to the Congressional Research Service, in 2017, there were:

Lawyers: 168 in the House of Representatives, 50 in the Senate.

Business folk: 179 in the House, 29 in the Senate.

Scientists? In the 435 member House we find 1 physicist, 1 microbiologist, 1 chemist and 7 engineers; in the senate there is exactly 1 engineer.

That's it, that's all of them.[8]

Let's look at a few who have earned the endorsement of 314 Action:

Louise Slaughter, U.S. rep for New York's 25th District. "...a trained microbiologist...As a member of the House Budget Committee, Slaughter was responsible for securing the first $500 million for breast cancer research from the National Institutes of Health."

As Rep. Slaughter puts it, "At a time when science and facts are under unprecedented assault, the work 314 Action is doing to recruit more scientists to run for office is more important than ever...We've got a country to save, which is why the time has come for scientists to step out of the lab and run for public office."

---

[6] https://www.nytimes.com/2017/08/07/climate/climate-change-drastic-warming-trump.html

[7] http://www.motherjones.com/environment/2017/01/house-science-committee-lamar-smith-climate-change-exxon/

[8] https://fas.org/sgp/crs/misc/R44762.pdf

Or Jerry McNerney, U.S. Rep for California's 9th District, a strong stem cell research supporter I have known personally — and a renewable energy expert who worked 20 years in the private sector developing renewable energy technologies...In Congress he authored...major legislation to help train the next generation for green jobs...

What if Jerry hadn't run? Richard Pombo, the man Jerry defeated, was considered one of the most anti-science members of Congress...

"Warner Chabor, the League of Conservation Voter's CEO, stated, "Having Pombo represent a district that includes Yosemite National Park is like electing Godzilla as mayor of Tokyo."[9]

Or Jacky Rosen, candidate for U.S. Senate in Nevada. As President of her synagogue, she "led a team to construct one of the largest solar arrays in her community, which reduced their energy bills by over 70%... allowing more money to go into...programs that provide food and housing for the homeless..."

But are scientists interested in exchanging a quiet lab for the roar of politics?

So far, "roughly six thousand scientists" have expressed interest in learning more about running for office, said 314's Ted Bortelon in a telephone interview.

"We have had two trainings already, and more are scheduled soon."

It will be hard for them. Never before have the anti-science forces been so free to contribute vast sums of money to the electoral process.

But as Ms. Naughton puts it, "Scientists won't win unless they run."[10]

For those who say, science is above politics and scientists should stay in the lab, she responds:

"How is that working for you?"

In a recent article, she summed up the problem:

"Donald Trump has supplied science skeptics with a megaphone. Radio host and anti-vaccine propagandist Robert F. Kennedy Jr. has been approved to chair a commission on 'vaccine safety and scientific integrity.' Scott Pruitt, a lawyer who made his career suing the EPA on behalf of oil interests, now runs it. Former Texas Governor Rick Perry is heading the

---

Department of Energy, the agency he pledged to eliminate — during his 2012 presidential bid."[11]

If those who know and respect science are subtracted from the political process, who are we left with?

For more information, visit the 314 Action group's headquarters: https://www.314action.org/mission-1/

---

[11] https://www.forbes.com/sites/jamesconca/2016/12/14/oops-rick-perry-chosen-to-head-energy-department/#6714a74d5be3

45 The Strangest Thing Inside My Head

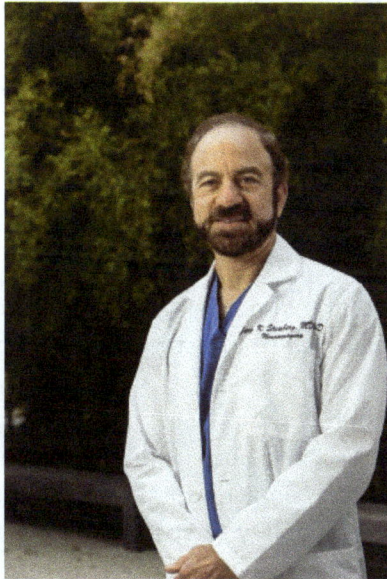

Dr. Gary Steinberg is attempting to defeat stroke, the number one cause of disability on earth. (Stanford.edu photo.)

In the middle of the night, I was wakened by a soft "pop" inside my head, like the sounds of bubbles bursting.

I could actually <u>see</u> the letters of the words in my thoughts.

Like: "I t-h-i-n-k s-o-m-e-t-h-i-n-g i-s w-r-o-n-g i-n-s-i-d-e m-y h-e-a-d…"

I poked Gloria beside me, and she woke instantly, lightest sleeper in the world.

"What, what?" she said, and I told her what was happening. It sounded odd, like describing a dream.

"Are you feeling any pain?" she asked, and I said no.

"Then take an aspirin and go back to sleep," she said, "If there was something wrong, it would probably hurt."

And so I did.

In retrospect, the aspirin was good advice, probably helping to thin the blood and lessen brain damage — but I should have gone to the emergency room.

I had just had a minor stroke.

Ever since, I have trouble with short-term memory loss. Nothing crippling, I can still do my job, and write (as I am doing right now!) but there is a difference. (The visualization of letters only lasted for a couple days.)

If Gloria tells me three things to do, I will probably remember the last one she said, like, "...and take out the garbage." The first two chores are gone. And sometimes I have to struggle to get words for common items, like "hand me the...the...the...pencil!"

It is not a big deal. At 74 years, a certain amount of memory loss is expected.

But had I gone to the emergency room immediately, and been treated with what is called a "clot buster", I might have skipped damage altogether.

Even so, I got off lightly.

Stroke is the number one cause of disability on Earth, the second highest cause of dementia, and the third highest cause of death. It can paralyze limbs, steal one's ability to speak, damage the mind, or take a life.

The CIRM "Stroke Fact Sheet" says:

"In the U.S., 795,000 people have a stroke each year; 140,000 die of it ..."

"Stroke is a chronic condition. Around the world, an estimated seventeen million people require long-term treatment, assistance in daily life, at great expense. Cost estimates run as high as $38 billion a year in America alone."[1]

How can you tell if you are having a stroke?

"<u>FAST — **F**ace droop, **A**rm weakness, **S**peech difficulties — Time to call 911!</u>"[2]

---

[1] https://www.cdc.gov/stroke/facts.htm
[2] https://www.saebo.com/rising-cost-stroke-america/

Children are by no means immune. 34% of all strokes happen to folks under 65 years of age. Sickle cell anemia is a frequent cause of stroke among teenagers. And once "stroked," sufferers remain damaged.

But CIRM-funded scientists are working to find a cure.

People like Stanford's Dr. Gary Steinberg, who recently completed an 18-patient human clinical trial.[2]

Drilling a hole in the patient's skull, he made a nickel-sized opening to the brain. Then he injected stem cells (mesenchymals, from bone marrow) which secrete growth hormone factors.

Within a few weeks, the stem cells disappeared. But their effects remained. For several patients (not all), the results were spectacular.

The day after treatment, one patient could lift her arms and legs. Her speech was near normal. "It was like I had been asleep and was now waking."[3]

Not every result is so spectacular. But in almost every instance, even in the safety-only procedures, when the lowest dose amounts of stem cells were injected, there was at least some improvement.

Importantly, this could benefit people with chronic (long-term) injuries.

For stroke, the window of opportunity is short. When I had my stroke, I did not know I had about four hours to do something about it.

If I had gone in immediately, something called a thrombolytic or "clot-buster" could have been injected into the big artery at the groin. This might have melted the blood clot interfering with my brain.

But the hours passed, the damage was done, and the window of opportunity closed.

Or maybe not?

Dr. Steinberg is attempting to heal people with chronic injuries, from six months to five years after stroke. And he is getting results.

"Since chronic stroke patients generally reach maximal recovery by six months, we previously thought the (stroke-affected nerves) were "dead" or irreversibly injured after that time. We now know that it is possible to resurrect these circuits...in certain stroke patients, even years after the stroke."[4]

What was it like to be a patient to receive such benefits?

---

[3] https://med.stanford.edu/news/all-news/2016/06/stem-cells-shown-safe-beneficial-for-chronic-stroke-patients.html

[4] http://stroke.ahajournals.org/content/qa-dr-steinberg

"I just started crying," (said stroke patient Sonia Coontz). When she tried to move her formerly paralyzed arm, instead of moving it inches, she raised it over her head. She tried her leg, and discovered she was able to lift and hold it up. "I felt like everything (had been) dead: my arm, my leg, my brain...and it woke up."[5]

You have heard the old saying, "This is not brain surgery"? Well, this is brain surgery: incredibly complicated and difficult.

But at the Stanford Stroke Center, (which he co-founded) Steinberg is working with champion stroke specialists like Greg Albers, Michael Marks, Neil Swartz, Maarten Lansberg, Heng Zhao, Paul George, Peter Tass, Marion S. Buckwalter, and others to develop a cure for stroke.

We do not want Dr. Steinberg's research to be merely promising steps forward — we want him and his colleagues to win outright, developing a reliable new standard of treatment — so when people have a stroke, there will be something they can do about it.

For that the scientists need reliable funding.

Another reason why the California stem cell research must go forward.

---

[5] https://www.smithsonianmag.com/science-nature/neurosurgeon-remarkable-plan-stroke-victims-stem-cells-180967211/

46 Of Crocodiles, and Politics

Gustave, giant crocodile of Burundi, was reportedly captured, just days ago, after killing an estimated 300 people .... (Photo by Wikipedia.com.)

There is a beast in Africa, a giant crocodile named Gustave. It has reportedly killed full-grown water buffalo, hippopotami — and some 300 people.

Photographs show Gustave as a legitimate monster, twenty feet long (the length of an orca) and weighing upwards of a ton. Villagers who live near Gustave appear to accept his presence (some attribute magical powers to him) and swim, fish and do their laundry in the water every day, never knowing if he is near.

Numerous attempts have been made to kill or capture him, but the enormous croc has always escaped, though identifiable now with bullet marks on head and side.

Why has he been uncatchable for decades? He has the crocodilian ability to burrow under a mudbank and "hibernate" there for months, perhaps years. He is also a great traveler, and migrates as he wills.

One major problem in catching Gustave is the ongoing civil war between the Tutsis and the Hutus. Crocodile hunting is difficult enough, without being shot at.

Similarly, the attempt to find cures is slowed by divisive politics.

Politics is inevitable, so long as people can disagree. If one party intends to reduce taxes on the rich, while the other wants to increase them, you have a guaranteed conflict. To an extent, the back-and-forth of politics is healthy. Both sides have something valuable to say; each should be heard.

But as the African war limits the hunt for Gustave, so politics interferes with our chances of defeating the $3 trillion dollar annual cost of chronic disease.

People like me are part of the problem. Republicans irritate me, and I find it difficult to cooperate with them.

Let me illustrate: Gloria and I were driving through our small home town in Northern California, when I noticed the car in front of me had a bunch of pro-Republican stickers cheering for various GOP candidates.

My breathing changed, and I muttered something. Gloria said, "Mmhmm," and put one hand to her mouth. Yawning, no doubt.

The car in front of me turned right, left and right again — as did I — it was in my neighborhood, heading down my street.

I followed — and that car turned left — into my driveway!

I leaped from my Honda, ran to confront the invader of my space.

Then my father got out of his car.

"Oh, uh, hi, Dad!" I said to 96-year-old Dr. Charles H. Reed, former Superintendent of Schools for Mt. Eden School District, and an ardent, life-long conservative Republican...

I don't talk politics with him, ever, but we do discuss stem cells. He is extremely religious, I do not share his opinions there, but listen to him politely, more or less.

If politicians could work together similarly, maybe we could cure paralysis and other currently incurable conditions.

Example: human embryonic stem cell (hESC) research.

In general, Americans support it 2–1 (65%) according to a 2014 Gallup poll.[1]

Regular Republicans? In 2007, it was in the news a lot because of President Bush's two vetos. Roughly 50% of Republicans support it. Independents give it a 63% thumbs up, and Dems are behind it by a massive 77%.[2]

But the leadership of the Republicans? They wanted (still do) to ban hESC completely. Look at their 2016 platform: They use the word "oppose" three times in their paragraph on stem cell research:

"We oppose embryonic stem cell research. We oppose federal funding of embryonic stem cell research…We oppose federal funding for harvesting embryos…"[3]

Do our differences get in the way? Amazingly so.

As detailed in my first book, "STEM CELL BATTLES", House Resolution 810 was a very modest national bill, simply allowing for the possibility (no actual money was dedicated to it) of funding embryonic stem cell research. No embryos would be created; strictly blastocysts (microscopic sperm–egg combinations) would be used, and only those already scheduled to be thrown away.

What was the vote? In the House, 50 Republicans supported it while 180 opposed. Among Democrats, 187 supported, with only 14 against.

In the Senate, 18 Republicans said "Yes!" while 36 said "No!" Democratic Senators voted 44 in favor, with only 1 in opposition.

We passed the bill in both House and Senate — but Republican President George W. Bush vetoed it twice.

So, if you want full stem cell research, there is only one party for you.

If you live in a state controlled by the conservative GOP, your chances of benefiting from embryonic stem cell research are very much reduced.

If you live in California or another Democratic state, you have a better chance at full research funding and freedom.

---

[1] https://news.gallup.com/poll/170789/new-record-highs-moral-acceptability.aspx

[2] https://news.gallup.com/poll/27898/six-americans-favor-easing-restrictions-stem-cell-research.aspx

[3] https://www.researchamerica.org/news-events/news/republican-platform-blasts-fda-seeks-embryonic-stem-cell-ban

Fortunately, this is not to say Republicans cannot come through sometimes! For instance, here is a lovely moment that just happened. A bipartisan budget was announced, which contained reasonable amounts of funding for research.

A non-partisan group called Research! America had this to say:

"Research! America is deeply grateful for the leadership and commitment from both parties that is reflected in this budget agreement for fiscal years 2020 and 2021. Research!America applauds this bipartisan, bicameral budget deal."

"By avoiding sequester and providing increased funding for Defense and non-Defense spending, this agreement empowers our nation to boost funding for the National Institutes of Health, the Centers for Disease Control and Prevention, the Food and Drug Administration, the Agency for Healthcare Research and Quality, the National Science Foundation, and other federal agencies that leverage research and science to save lives. This truly benefits all the American people."[4]

I should learn from them!

Gustave, unfortunately, cannot.

---

[4] https://www.researchamerica.org/news-events/news/statement-researchamerica-president-and-ceo-mary-woolley-bipartisan-budget-act-2019

47 Raja's Story

Ovarian cancer attacked my friend Raja.

If you read one of my earlier books, "CALIFORNIA CURES," you might remember my good-hearted barber friend, Raja, originally from Thailand.

Lean and muscular, Raja rode her bike to and from work every day, six miles each way. She was also a champion bowler.

The smiling part of Raja's story had to do with her advice on baldness, which so far has passed me by, apparently because of her helpful tip on aloe vera. I massage the cactus-based clear gel into my scalp every morning; and at 74, my head still has most of its foliage. When I stop using the gel, my shower drain plugs up with hair right away. When I go back on the gel, the hair stops falling out.

But Raja's personal story begins and ends in sadness.

She had a terrible childhood in Thailand, essentially being sold as a slave to a neighbor because her family was starving. Her father had died, her mother worked two jobs, and there was never money for Raja to go to school. So, she became a worker at age 8. Other cruelties were hinted at, but not disclosed.

But her mother wanted her to have a chance, and scraped together savings for years to send Raja to America. There she worked for a cousin. Time passed, and Raja was frugal, and ended up with her own business: a barber shop, for which she was the sole employee. She worked 7 days a week. Her shop essentially never closed.

She adopted two children who lived in Thailand. She paid for their living and job training, and one became a nurse and the other a policeman. She kept every report card mailed to her, and corresponded, having taught herself to read and write.

She took in homeless animals, feeding dogs and feral cats behind the shop.

And then one day she was diagnosed with stage 4 ovarian cancer.

I knew a scientist friend working on that condition, who had a clinical trial coming up. I looked up the trial on www.clinicaltrials.gov, and it still said "recruiting" and there was no fee.

I contacted the scientist, who said if she fit the protocol, he would be glad to accept her in the trial. Getting qualified participants was not always easy, he said.

All Raja had to do was fulfill the requirements (mainly undergoing the same treatments she would have done if there was no stem cell treatment available) after which she could become a patient. The scientist gave me contact info for the people she had to deal with, as well as paperwork for her current ob/gyn doctor and the contact info for the person in charge of the trials.

She would need chemo, and probably surgery, to try and remove the ovarian cancer. If that worked, there was no need to go further. If not, then there was nothing at risk, because the standard treatments would have been tried.

I wrote down everything she had to do, also explained it clearly as I could.

I gave her the following note: names removed.

Dear Raja:

I would like you to consider being part of a very advanced stem cell clinical therapy, developed by Dr. _____ of _____.

1. If interested in being part of _____'s stem cell therapy, you need to first have your doctor (ob-gyn) talk to Dr. _____ at _____. His email is _____.
2. Then you will need to have standard treatment.
3. The experimental treatment will take place in _____.
4. If you want to talk to me about it at any time, I can be reached at _____.
5. There will be no financial charge to you at all.

Please think carefully about this, Raja. I lost two family members to cancer, and this is what I would want them to do, if they were alive.

I talked to the scientist doing the work, and everything seemed to be all set — but Raja seemed nervous, as if she was putting it off — and finally told me she did not want to do any of it, not even the chemo or the surgery.

"No worries, I be fine, thank you very much," she said. We hugged goodbye.

And then one day when I came for my haircut, the store was closed.

Several months later, it re-opened, with another barber, who had bought the shop.

Raja had gone home to Thailand, said her friend.

There she saw her mother for a little while, spending time together, after which they both died.

Why am I telling you this sadness?

Raja and I were friends, and the cancer killed her, as it has done to other members of my family.

You have a friend like Raja, too, maybe.

We should honor his or her memory — by defeating cancer in our lifetime.

48 To More Swiftly Heal a Broken Bone?

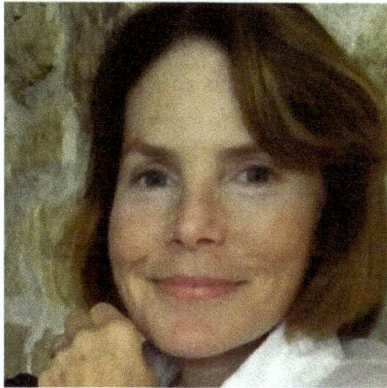

Dr. Jill Helms of Stanford may have found a way to accelerate the speed of bone healing by as much as 350%. (Photo by Stanford.edu.)

It was 4:00 in the morning, March 13th, 2019. I was at Roman's house, providing his routine medical care.

The doorbell rang. Who could be visiting at such a miserable hour?

It was my brother-in-law, Salvatore LaManna.

"Gloria's fallen," he said, "The fire department is already there."

Early-riser Gloria had been carrying a basket of laundry outside, to hang it on the clothesline.

Her knee gave out on the right side; she slipped and fell. The impact exploded the basket into orange plastic fragments. Fortunately, her face landed in the laundry, which padded her fall, preventing a possible skull fracture.

Her arm and shoulder hurt terribly, and she could not get up.

She might have lain there in the dark for hours, if not for neighbor Lisa A., who heard Gloria's groans and came to her assistance, contacting the fire department...

When I reached home, Gloria was arguing with the firemen. They were saying they could not leave her without help, and they needed to call an ambulance for her. And Gloria was saying oh, no, that was way too expensive.

On our drive to the hospital it seemed like every pot-hole in California was reaching for our car. Every bump was a jolt of pain for her.

At the hospital they x-rayed her arm and shoulder, and set up an IV drip of morphine into the back of her wrist.

It was long hours before the doctor gave his diagnosis.

"Your shoulder is broken; we're sending you home," he said.

That did not make sense. If she was injured, should she not stay in the hospital?

"No," said the doctor, "This kind of bone injury might hopefully heal on its own."

They put her on pain medication: Oxycodone tablets, one every six hours.

"I am going to be addicted," worried Gloria, and her concern was not without foundation. Oxycodone is one of the most addictive drugs on the market.

But what else could we do? The pain was incapacitating, and even the pills just took the edge off, from a 10 to at best an 8, or maybe a 9.

Until she is well, I must be her arms and legs and hands, the extension of her will.

It must be comical to see her ordering me around the kitchen, dictating my every move. My cooking skills, to put it kindly, are limited. On my own, I would live on pizza, smoothies and cold cereal. But Gloria intends for us to eat properly, and since she cannot cook right now..."On the right-hand corner of the third shelf are two bottles of spices, one on top of the other; you need one shake of each..."

And of course there is dressing, bathing, house cleaning, all the endless chores we take for granted until we cannot do them.

How long would it take for her fractured shoulder to heal? Several months, probably.

Naturally, I went to the California stem cell agency's website, to see if there were any clinical trials going on which might help.

To my delight, I learned that CIRM has been funding a bone repair therapy: to accelerate the healing of aged (as in senior citizens') bones.[1]

Dr. Jill Helms of Stanford had been experimenting with WNT-3A, a protein. It accelerates growth for certain body tissues, including bone.

In her experiment, anesthetized mice had tiny holes drilled into their legbones. When they woke up, they were put on pain-killers, and injected with WNT-3A.

With WNT-3A, their *healing accelerated* as much as 350% — more than three times as fast as a non-treated mouse. The WNT-3A mice were completely healed after 28 days, unlike the control group (no growth factor) which took much longer.[2]

And this Thursday, March 21, an advanced form of WNT-3 bone healing therapy will be judged by the board of directors (ICOC, the Independent Citizens Oversight Committee) of the California stem cell agency.

The proposal was CLIN 1-11256, for almost $4 million ($3,994,246), of which $998,562 would be co-funding.[3]

The project would not help Gloria, being narrowly focused: to repair backbones of patients having spinal fusion surgery.

But if it succeeded, it would be a huge step in the right direction, establishing procedures for a clinical trial, including:

Trying out a toxicology test in a rabbit model of spinal fusion, mimicking the surgery to be done in patients.

Setting up the GMP (Good Manufacturing Practice) of the WNT-3A protein, suitable for use in human patients.

Preparing an Investigational New Drug filing with the Food and Drug Administration (FDA).

Long-term, its potential seemed extraordinary. If WNT-3A could fix one kind of broken-bone injury, why not others?

This year, 6.8 million Americans will suffer bone fracture. Many will be of an advanced age, like Gloria.[4]

---

[1] https://www.cirm.ca.gov/blog/05052010/protein-aids-bone-healing

[2] https://www.cirm.ca.gov/our-progress/awards/enhancing-healing-wnt-protein-mediated-activation-endogenous-stem-cells

[3] https://www.cirm.ca.gov/sites/default/files/files/agenda/CLIN1-11256%20ICOC%20Summary.pdf

[4] https://www.schwebel.com/userfiles/files/Fractures(1024).pdf

We must find a better way to heal them — friends, neighbors, family, loved ones — not just send them home with pain meds, hoping the injury will heal on its own.

P.S. Naturally, I wrote a letter of support for this important effort, which would be decided at the next meeting of the board: the Independent Citizens Oversight Committee (ICOC). Maria Bonneville of CIRM was kind enough to read it into the public record.

Dr. Helms was in the room, and shared the following note:

"DON!!!!!!! Oh my gosh, I am so excited and relieved! Maria Bonneville read your statement and the room was so quiet, you could hear a pin drop. It was a poignant and important reminder to all of us why we are doing this...

"My best to Gloria..."

— Jill Helms (personal communication).

And then there was nothing to do but wait, as one by one the members of the board were asked for their decision.

"Jeff Sheehy?" "Yes."

"Senator Art Torres?" "Aye."

"Francisco Prieto?" "Aye."

"Lauren Miller?" "Yes."

"Dianne Winokur?" "Yes."

"Steve Juulsgaard?" "Yes."

"David Higgins?" "Yes."

"Joe Panetta?" "Yes."

"Adrianna Padilla?" "Yes."

"Anne-Marie Duliege?" "Yes."

"Dave Martin?" "Yes."

And so it went.

Five minutes later, the calmly-spoken crucial words: "The measure passes."

Three words, that mean so much! The measure passes...

Nearly four million dollars will be dedicated towards a more rapid and effective way of healing fractured bones. CIRM's press release summed it up:

"Ankasa Regenerative Therapeutics (Dr. Helms' company) has developed... a product meant to enhance the bone healing properties of...bone grafts. (It) works by stimulating bone stem cells..."

Does this mean in five or ten years, there will be a cure for bone fractures?

We cannot know. But if we do not try, we can absolutely guarantee there will be no progress, no successes, no cures.

And this is how CIRM does its job: supporting and coordinating the efforts of scientists, public education institutions, and private enterprise.

So long as the funding lasts, they will go on, fighting for a better way.

P.S. Unfortunately, Gloria's injury was more serious than we had hoped.

On April 3, 2019, she underwent full shoulder replacement surgery.

49 Surprises, Awkward and Otherwise

Stem Cell Research Center: CIRM cooperated with champion donors Sue and Bill Gross to make the magnificent stem cell research center at UC Irvine .... (Photo by Sue and Bill Gross.)

One night I dreamed three robbers invaded my home. Being naturally heroic in my dreams, I attacked them.

The fight grew more and more ferocious. I used my Wing Chun, short punches and low kicks to no avail. Finally I resorted to desperate measures. I grabbed the leader and performed a Gracie Brothers Brazilian jiu-jitsu shoulder throw on him.

I flipped the enemy so vigorously, I also flung myself out of bed.

The THUD! of landing awakened both myself and my wife.

It took a moment to absorb the situation. Let's see, I am on the floor in the middle of the night, on forearms and knees...There must be a logical explanation...

Gloria asked me, what was I doing?

I told her how I was battling home invaders, and had judo flipped the leader.

"That's nice," said Gloria, "I hope you won."

Then she turned over and went back to sleep.

A more useful surprise came to the people of California, by way of CIRM's infrastructure grants.

We hear a lot about infrastructure in the news; mainly road repair, paid for by gas taxes. In medical research, it can be buildings, labs, equipment, etc. — all of which has to be paid for.

But what if the infrastructure of medical research had a head start from the state — after which the generous-minded came through with big donations?

Bob Klein once explained to me that rich people, through their foundations, liked to support projects that would grow and last and attract financial muscle on their own — like CIRM and its centers of excellence — 12 new labs for scientists.

CIRM's major facilities program has created state-of-the art (facilities) for carrying out stem cell research. The agency's initial investment of $271 million leveraged $543 million in private donations and institutional commitments toward 12 (stem cell research) buildings...

John Pérez, California Assembly Speaker, said: "CIRM's investment in stem cell research buildings created jobs and revenue for the state and solidifies California's position as the leader in developing the stem cell therapies of the future."[1]

Here is what happened, when CIRM either started the ball rolling, or contributed to a process well begun — and others chipped in to a successful team effort.

Notice there are two figures cited below, underneath each institute. One is the total expenditure; the other is CIRM's contribution. "M" of course stands for million.

---

[1] https://www.cirm.ca.gov/our-impact/creating-infrastructure

And now these labs are contributing to the fight against chronic disease and disability.

In these awkward times, when the folks in Washington seem incapable of working together, CIRM's success is a pleasant surprise indeed.[2]

Buck Institute for Research on Aging
Regenerative Medicine Research Center
Total: $36.5 M
CIRM: $20.5 M

Sanford Consortium
Sanford Consortium for Regenerative Medicine
Total: $127 M
T. Denny Sanford: $19 M
CIRM: $43 M

Stanford University
Lorry I. Lokey Stem Cell Research Building
Total: $200 M
Lorry Lokey: $75 M
CIRM: $43.6 M

University of California, Berkeley
Li Ka Shing Center for Biomedical and Health Sciences
Total: $257 M
Li Ka Shing: $40 M
CIRM: $20 M

University of California, Davis
Institute for Regenerative Cures
Total: $62 M
CIRM: $20 M

University of California, Irvine
Sue and Bill Gross Stem Cell Research Center
Total: $80 M
CIRM: $27.2 M
Sue & Bill Gross: $10 M

---

[2] https://www.cirm.ca.gov/about-cirm/newsroom/press-releases/01152010/all-12-cirm-major-facility-projects-moving-forward

University of California, Los Angeles
Eli and Edythe Broad Center of Regenerative Medicine and Stem Cell Research
Total: $43 M
The Eli and Edythe Broad Foundation: $20 M
CIRM: $19.8 M

University of California, Merced
Stem Cell Instrumentation Foundry
Total: $7 M
Ed and Jeanne Kashian: $100,000
CIRM: $4.4 M

University of California, San Francisco
Ray and Dagmar Dolby Regeneration Medicine Building
Total: $123 M
CIRM: $34.9 M
Ray and Dagmar Dolby: $36 M
The Eli and Edythe Broad Foundation: $25 M

University of Southern California
Eli and Edythe Broad CIRM Center for Regenerative Medicine and Stem Cell Research
Total: $80 M
CIRM: $27 M
The Eli and Edythe Broad Foundation: $30 M

University of California, Santa Cruz
Institute for the Biology of Stem Cells
Total: $83.7 M
CIRM: $7.2 M

University of California, Santa Barbara
Center for Stem Cell Biology and Engineering
Total: $6.4 M
CIRM: $3.1 M

50 The Man with the Plan to Assassinate Cancer?

At UCSD, Dan Kaufman is using Natural Killer Cells — to try and assassinate cancer! (Twitter.com photo.)

It is said that when a person dies, his or her brain offers one final gift; highlights of its owner's past. Favorite moments return, and images of people he or she loved.

If that is true, I hope to see once more my sister Patty, who died of leukemia. I miss her every day and wish so much my grandchildren could have known her.

I wish I could physically attack the disease that took her life. That of course would be difficult.

But the best revenge would be to defeat that miserable liquid cancer, so other families might be spared the agony of loss.

The California Institute for Regenerative Medicine (CIRM) is working to end leukemia, and every form of deadly cancer.

Thanks to CIRM, I have had the privilege of meeting some of the giants in cancer research: people like Stanford's Irv Weissman, the bearded colossus who discovered the cancer stem cell, evil twin of the good stem cell. His work explains remission: why cancer may seem to be gone — and then return — because the cancer stem cell can hide. Thanks to the Weissman research, the "cloak of invisibility" can be lifted away, exposing the cancer stem cell, giving the body a chance to attack it.

A smiling warrior is Catriona Jamieson, Deputy Director of UC San Diego Moores' Cancer Center Called "Cat" by her friends, Dr. J. works tirelessly to defeat not only leukemia, but conditions leading to it, like polycythemia vera.

But when I read the research of Dan Kaufman, I was stunned. A photograph of three mice was hideous to look at, their bodies riddled with cancer. Beside it was another picture of three rodents, their white coats shimmering, radiant with health — it took a moment to realize that these were *the same animals*...

They had been healed of their cancer. They were not just better, they were well.

Could this be the long-sought answer to cancer?

What a possibility! Think in terms of just money for a moment.

How much does blood cancer (such as leukemia) cost America?

An average $112,000 — for one patient, one year.[1]

The total number of people living with leukemia? According to the Leukemia Research Foundation, 387,000.[2]

$112,000 times 387,000 people with leukemia = $43.3 *billion* dollars — every year.

Kaufman's proposed therapy is designed to attack solid forms of cancer as well...

How does it work? Dr. Kaufman began with Natural Killer (NK) cells, a white blood cell, part of the immune system.

---

[1] http://tinyurl.com/y257namc
[2] https://www.allbloodcancers.org/disease-information-and-support-statistics

Something called a Chimeric Antigen Receptor (CAR) is added to the NK cells, where it functions like glue. The resultant cell (now called a CAR-NK, or iPS-NK) can stick to a protein on the surface of the cancer — and kill it.

As Dr. Kaufman says:

"We've shown you can engineer...Chimeric Antigen Receptor-expressing NK (CAR-NK) cells to better target refractory cancers that have resisted other treatments."[3]

Large numbers of CAR-NK cells can be made, and stored, ready to be used. They need not be taken from each individual patient's body, as an exact match is not required. This is a great time saver, crucial when fighting cancer. When cancer spreads throughout your body — you want relief right now — not in several months!

An off-the-shelf cancer-killing medication? Wow...

Using CAR-NK cells may one day be as simple (and spectacular) as going to the doctor's office, getting a shot, and the cancer goes away...

To develop the therapy, Kaufman is working with a company called FATE THERAPEUTICS, a name to remember.[4]

How close are we, in terms of testing IPS-derived natural killer cells (NKs) on people?

Right now, at the University of California at San Diego, a man named Derek Ruff has Stage Four colon cancer. How bad is that? There is no stage Five. Stage Four means the cancer has spread to distant organs in his body. The disease had been in remission for ten years, and then it came back.[5]

Mr. Ruff has been given some of Kaufman and Fate Therapeutic's off-the-shelf, iPS-derived ready to-go NK stem cell treatment. We do not yet know the results.

I am sure I speak for every family with a loved one threatened by cancer, when I wish Derek Ruff complete success in his trial: may the cancer by wiped away, so that he regains complete and lasting health.

---

[3] https://ucsdnews.ucsd.edu/pressrelease/car_t_immunotherapies_may_have_a_new_player

[4] https://fatetherapeutics.com/collaborations/

[5] https://blog.cirm.ca.gov/2019/04/10/first-patient-treated-for-colon-cancer-using-reprogrammed-adult-cells/

Our hearts are with you, Mr. Ruff. You are fighting not just for yourself, but for all of us. Godspeed!

What lies ahead, as Kaufman works to develop a reliable and reproducible cure?

"More trials, lots of activity!" he told me in a recent conversation. As always (I have known him for years, since before he moved here from Minnesota) he sounded tough, cheerful, and determined — ready to face whatever challenges arise.

The world is waiting… for the man with the plan to assassinate cancer!

51 Lung Cancer, and the Bent Cigarette

Humphrey Bogart was called the "King of Movie Stars," until cigarette smoke ended his career — and his life. (https://en.wikipedia.org/wiki/Humphrey_Bogart)

My smoking career began at the age of 12, when my older sister Patty and I were watching a Humphrey Bogart movie.

"Doesn't he look cool, how he smokes?" she said.

Now you have to understand, anything Patty said to me was like engraved in letters of gold in my brain. If she thought smoking was cool...

I found an old pack of cigarettes some visitor had left. They had been sat on, and were mildly crushed, but still seemed reasonably functional.

I lit the filter end first, which burst into flames. This did not seem quite right, so I tried again. The next attempts went better, aside from a near-strangling cough.

At last I was ready. I told Patty to come out into the back yard, I wanted to show her something.

I straightened out one of the less-damaged cigarettes, lit the proper end, inhaled into my mouth so I would not choke, exhaled — everything was going perfectly —

Patty burst out laughing.

"Oh, I am sorry," she said regaining her composure, "It just looked so funny…"

I threw away the bent cigarettes.

Lung cancer is the deadliest form of cancer: killing more people than breast cancer, colon cancer and prostate cancer combined. Cigarette smoking is the number one risk factor for lung cancer.

Smoking is also an important risk factor for Idiopathic Pulmonary Fibrosis (IPF), a fatal stiffening of the lung tissues. Generally, a person diagnosed with IPF will last no more than 3–5 years.

Dr. Brigitte Gomperts of the Broad Institute at UCLA is fighting both.

As colon cancer can arise from polyps in the intestines, lung cancer is thought to come from lesions in the airways. Studying these lesions appears to be critical for understanding how lung cancers form.[1]

Lesions are wounds, which can be caused by cigarette smoke. My first choking inhalations were actual injuries, though fortunately I stopped short of lasting harm.

Gomperts' research focuses on the role stem cells play in repairing and regenerating the lungs after injury, including smoke. She studies how this repair process can go awry, which can lead to cancer, idiopathic pulmonary fibrosis and other lung diseases.[2]

By studying the lesions, her team has identified molecular changes in the stem cells that can lead to lung cancer.[3]

Her studies led to a strange conclusion: part of the problem of lung cancer was caused by a malfunction in the body's *repair system*.

---

[1] http://tinyurl.com/y5fbgus8

[2] https://stemcell.ucla.edu/member/gomperts

[3] Ooi AT, Gower AC, Zhang KX, *et al*. (2014) Molecular profiling of premalignant lesions in lung squamous cell carcinomas identifies mechanisms involved in stepwise carcinogenesis. *Cancer Prev Res* 7(5); 487–95.

That seems like hiring a mechanic, and he/she takes a crowbar and starts smashing your engine!

By understanding how the body's repair system goes bad, new therapies may be identified that promote healthy airway tissue and prevent lung cancer.

Apparently there are metabolic changes in the airway stem cells, and these changes can be manipulated to alter the cells' behavior. For example, there are two metabolic proteins in the repair system (SLC2A1 and LDHA) and if they get too active, they may actually begin cancer and contribute to its growth.

If those proteins are calmed down, the problem may be resolved. Cancer cells not previously sensitive to chemotherapy may lose this protection and be killed.

In scientific lingo:

"SLC2A1 and LDHA inhibitors, when administered in combination with other chemotherapeutic agents, showed synergistic antitumor effects by re-sensitizing chemo-resistant cancer cells to the chemotherapies. These results indicate that disrupting SLC2A1, LDHA of other regulators in cancer cell energetics is a very promising approach for new targeted therapies."[4]

See how simple? Yeah, right. I had to study several of Dr. G's papers over and over, about ten times, to understand even a little. Reading scientific papers is for me like when TV character Penny tries to understand genius Sheldon on the Big Bang Theory. But it has to be done. Public financing of science will not happen if the public does not understand it.

Another problem had to be resolved: scars. It was not enough to slow or stop the cancer; damaged tissue in the lungs must be replaced or repaired. This was a major problem in IPF as well.

To deal with this difficulty, Dr. Gomperts and her colleagues invented an amazing tool, the organoid, which functions like a miniature lung.[5]

They make tiny balls of alginate (made from brown algae) and coat it with cells from a human lung. Enough of these can add up to a miniature

---

[4] Ooi AT and Gomperts BN (2015) Molecular Pathways: Targeting Cellular Energy Metabolism in Cancer via Inhibition of SLC2A1 and LDHA. *Clin Cancer Res*: 21(11)2440–4.

[5] https://stemcell.ucla.edu/news/researchers-use-stem-cells-grow-mini-3-d-lung-dish

lung. This can even be engineered to grow the deadly scars, so you have a living tissue model of IPF.

How did Dr. Gomperts get started fixing lungs? According to her bio:

"Brigitte Gomperts, M.D., treats young children with cancer and blood diseases. Observing the side effects that chemotherapy, radiation therapy and bone marrow transplants had on her patients' lungs, she became interested in lung repair and regeneration."[6]

Is that not beautiful? Helping children inspired her to fight lung disease and cancer...with some of her work funded by the California stem cell agency.

And Humphrey Bogart? Unfortunately, America's greatest movie star did not have an older sister to warn him of the dangers of cigarettes.

A heavy chain smoker, Bogart died at the age of 57, of esophageal cancer.[7]

---

[6] https://stemcell.ucla.edu/member/gomperts
[7] https://oralcancerfoundation.org/people/arts-entertainment/humphrey-bogart/

52 Introducing Madame President

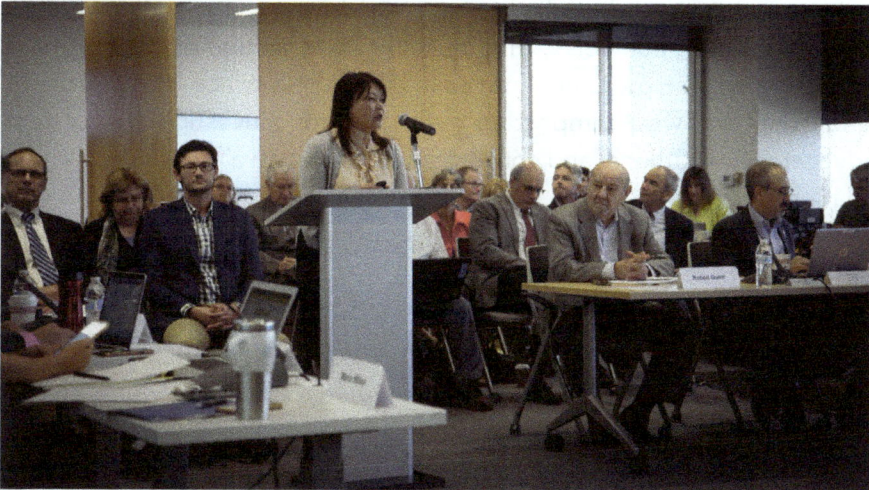

CIRM's new President, Maria Millan takes office. (CIRM photo.)

Kevin McCormack: "And now, two words I have been longing to say: I am delighted to introduce — Madame President!"

He was referring, of course, to Dr. Maria Millan, new President of the California Institute for Regenerative Medicine (interviewed below).

When Harry Truman became President, he said he felt like the moon, the stars and all the planets had fallen on him; what did it feel like to become President of CIRM?

I remember we were rolling out the new funding structure. I was working as head of therapeutics and infrastructure, and very excited about the alpha clinics. We were bringing in high quality clinical trials — with a goal of having 50 trials. At first that seemed like a too-bold goal. Some experts had thought we would be lucky to have one! But we had built a relationship with high quality scientific investigators, and our team was really hitting its stride in the first 2 years of the strategic plan. I was all excited about the accomplishments; the clinical trials were enrolling patients. We were making things happen.

But then one day Randy (Mills, then CIRM's President) took me aside, and said we needed a new president — and he thought it should be me. I was surprised that he would want to leave so early, but he said, CIRM needs new leadership now. I was in shock, because we were on a trajectory. But what I also felt was that we had so much to do: that whether it was me or someone else, the Presidency was a huge responsibility. It had to be done. How could I not take this on? I was accepting a role of leadership that would impact so many: scientific investigators, patient advocates, institutions, and corporate leaders. I had a great professional relationship with Randy, so it meant losing a colleague, but the job was amazing and it had to be done. We needed to go ahead, full force!

As President, what do you spend most of your time on?

Reporting back to the public as well as keeping in touch with other scientists is vital. As our projects move forward to the clinic, we must stay connected with the public, keeping Californians aware of our progress, advancements and challenges. How can we expect the public to support something it does not understand? We must also stay connected with the progress in science. As we achieve more FDA-approved products, we must keep track of where the world is going.

We need to share technology and data science and keep up with advances in all curative sectors. Standing still is not an option. I try to meet with key opinion leaders in all data sectors. Knowing the science goals and achievements of those outside CIRM will play a part in how we structure our own funding and partnering. I try to be aware of and responsive to the entire research ecosystem. Any advance in technology, even in something seemingly unrelated (solar panels, for example, may cut lab electrical costs) which might affect our progress. In order for us to serve the public, we must make our therapies accessible. The science is complex. It needs up-front investment to enable later cost savings. As initial treatments become effective, we must seek to achieve lower costs. So many things must be done; insurers need to understand the value story, so we need a large data exchange. Some components we have in place already, others we need to develop. We must learn how to share important information, so we can best inform patients, and how to cover them. Many are not yet aware of our program and how it might affect them positively. We will need a healthcare economics analysis, to anticipate the business side of getting therapies to patients, making curative solutions available to all.

If California renews CIRM's funding, what would you most like to focus on? Overall direction?

We have five parts to our mission: infrastructure, for the buildings which house the laboratories; education, to share what we have learned; discovery funding, for the scientists to make great leaps into the unknown; translational science, where their discoveries are matched up to the needs of people; and clinical trials, where every product and therapy undergoes rigorous testing for safety and efficacy. We need funding to continue what we been doing — systematically proceeding to meet patients' needs. And now the field is growing. We will need more workforce development to insure we have enough trained technicians and scientists to meet the need. As we work toward production of therapeutics, we need to put infrastructure in place, to manufacture products in the most cost-effective way, optimizing access to patients here, across the country and around the world. Above all, we must never approach a product as if disconnected from patients. They are the ultimate consumers, and must be empowered with knowledge — to navigate the system and access the best treatments available.

What do people know/not know about CIRM — How do they react when you tell them about it?

Since I joined CIRM 6 years ago, there is much more visibility. Years ago, when we talked about stem cells, people immediately went to ethical issues. Now the narrative has changed. There is greater understanding of the agency's goals and identity. The public insists the science be done ethically at all times, as do we. But they tend to understand that we sometimes deal with microscopic cells (embryonic) which would otherwise be discarded, and also sometimes use cells made from the patient's skin samples. They are excited to hear how the science is developing new therapies, and that new programs are currently in human trials, targeting devastating diseases. Progress is being made, fighting eye diseases, immunodeficiency, cancer — even with conditions where we have as yet no cure, there is the prospect of solving these formidable diseases. This hits home, because we all know someone with a chronic disease. This is personal.

Does CIRM help the scientists if they run into problems along the way?

CIRM maintains a unique, collaborative partnership with the scientists we fund. Because the research is so new and cutting-edge, we know that challenges will arise, and we want our investigators to know that we are with them. We keep a running dialogue on their progress and problems. Both they and their projects have already been evaluated as high potential, and we want them to succeed. Even so, all of the projects are difficult. If it was easy, it would already have been done. Each project has an advisory panel, which meets regularly, offering assistance and advice. This involvement is critical, often helping scientists overcome hurdles.

Also, we establish well defined milestones and expectations. This is not done to be dictatorial, but for clarity. Patient advocates are an integral part of the process, providing frank input early on, valuable insight from the people most directly affected by the disease or disability being challenged. This may help overcome hurdles. I am proud to say other research programs are imitating our policy of patient advocate involvement. Why is this valuable? Suppose a scientist is having difficulties achieving enough patient involvement, because the process of the research is not clear. Patients trust other patients, because they have similar experiences, and can relate. For instance, some of the protocols in preparing for a bone marrow transplant have been very strenuous in the past, hair loss, etc. But a modern protocol may be far more gentle. A patient advocate can explain this, and be understood, and trusted — because he/she has direct personal experience.

What would most be lost, if California does not renew?

The loss of the California Institute for Regenerative Medicine would hurt science everywhere, because CIRM is a cooperative partnership of people and institutions. We work hard to be an honest broker, influenced only by the merits of the science. We want everyone to succeed; we are not backing just one horse in this race. As we try to problem-solve, we hope the various solutions will apply to problems in many areas. We are pleased to play that role — inspiring, aggregating and convening — bringing people together so everyone wins.

CIRM supports the efforts of many women scientists — what advice/ encouragement would you give to girls or women considering a career in science?

To all, regardless of gender, my advice is this: follow your passion, live a meaningful life. Do not be dissuaded from something you love. When you encounter hurdles, remember these are not always bad. They may point you in directions you might not otherwise have considered. Do not be discouraged by things you don't know. Get deep in your area, and look for others with expertise for missing pieces of the puzzle. Learn from the past, and contribute wherever you can. The old method of keeping knowledge hidden, the "silo" approach, is not productive. The challenges we face today are so complex that no one can have all the answers. CIRM was set up to share knowledge and we do — so please be a part of the sharing community. Find good folks with whom to partner, and stay humble, for there is always more to learn.

What do you feel proudest about CIRM?

I feel proud that I can talk to Bob Klein and all the hundreds of people who worked together to build CIRM. I can look them straight in the eye and say: look, how well it is doing. See what came of your vision! This is real, not a made-up dream. No one has to say, "Oh well, we tried hard, too bad it did not work out!" No, just the opposite. We can smile and say, "Look what has been done!" Yes, there is still more work to do — but the dream led us in the right direction. We can honestly say it was the right thing to do — and so much more than we anticipated.

What is the most fun part of your job?

Problem solving! My job allows me to stay in touch with experts every day — corporate folks, students, doctors and researchers — people who believe for every problem there is an answer. And on my desk is a reminder for me: a little box from my days as a surgeon. We always maintained records, and in that box are stickers with names of more than 1,000 people I operated on: men, women and children who needed kidney and liver transplants, tumor resections, reconstructive surgery and more. Each one of them came into the hospital with a problem, which we fought; CIRM is a continuation of that effort: to save lives and ease suffering.

53 Fighting Bladder Cancer

Bladder cancer is all too often a death sentence. CIRM-funded scientist Philip Arden Beachy is fighting for a cure.

Imagine a nightmare scenario. You go to the bathroom, and urinate — blood.

First, of course, you would wrack your brain for some harmless explanation: maybe you bumped into something and injured yourself, and that was why your kidneys were bleeding? Professional boxers sometimes lose blood after a fight, their bellies punched so many times; after the "Rumble in the Jungle" with George Foreman, Muhammed Ali reportedly bled for a week.

But finally, you would force yourself to go to the doctor, hoping desperately you would not hear the grim diagnosis:

"You have cancer of the bladder."

Untreated, bladder cancer almost invariably leads to early death.

Treatments are possible, though harsh: to remove the cancerous portion of the bladder; or, you might have a "radical cystectomy" and remove the entire bladder.

If you chose the removal, you might wear a bag called a *stoma*, to be emptied every few hours. Or, a substitute bladder can be built from a piece of your intestine, connected to the kidneys. This empties through a catheter, sticking out of your stomach. These are major changes, and (while doable) are not pleasant.

Worse, the cancer may return. Even when 99% of the cancerous tissue is removed, that little bit remaining is like a "reservoir" of cancer stem cells (the bad kind), which may reproduce, and take over the bladder again, threatening your life.

Recently I had lunch with Drs. Lay Teng Ang and Kyle Loh, who (along with their colleague Philip Beachy, the Primary Investigator) hope to use stem cells to grow a cancer-free lining of the bladder.

In their approach, a catheter tube will be inserted into the urethra, and stem cells injected (painlessly) into the tube.

"We propose a new stem cell…therapy as a definitive cure for bladder cancer. Over the past six years, we have invented methods to turn embryonic stem cells into bladder cells…" — Lay Teng Ang, Personal Communication.

To which Dr. Loh added:

"… we first propose to destroy the cancerous cells. Then, we will replace them with the normal, healthy cells generated from embryonic or induced Pluripotent Stem (iPS) cells." — Kyle Loh, personal communication."

But the greatest science in the world cannot work, if there is no funding for it.

We take you now to the California Institute for Regenerative Medicine headquarters, 1999 Harrison Street, Oakland, California.

They have complicated elevators. You need to go to the lobby desk, and a cheerful person (typically an Oakland resident, as CIRM makes an effort to hire as many local folk as possible) will ask you where you are going, give you a Visitor's pass, and tell you which elevator to use.

The elevator knows where you are going, which is both neat and kind of threatening, like the HAL computer in the classic sci-fi movie 2001: A Space Odyssey.

So you have the kind of short conversations possible on an elevator ride. "Hi, are you going to the California stem cell meeting today, which is of course open to the public?" And they say yes or no, and get off the elevator as quickly as possible.

By now you know I like to get involved if there is a way a non-scientist can help, and I really liked this bladder cancer project, even with its cumbersome name:

"Pluripotent stem cell-derived bladder epithelial progenitors for definitive cell replacement therapy of bladder cancer" — Application: DISC2-11105.[1]

The immediate problem, of course, was the money was almost gone. There was $866,000 left in their particular category — basic science — and about ten groups vying for that money. So — a ten million dollar need, and less than a million to pay.

Things got worse. We arrived early and read the handouts (always a good idea), which said: for basic science, the funding was gone. Zero dollars were left.

CIRM was designed to have three parts, divided by time: first was basic (sometimes called discovery) science, then translational, and finally clinical. This was deliberate, and logical. First, CIRM was helping build a new field, and everything needed was basic science: lots of Petri dishes, pipettes and test tubes. Second came translational, where the early experiments were carefully molded into projects with the possibility of helping people; and finally, the clinical phase, where projects were actually tested on volunteers: people with the disease who wanted to be in a test, to find out first if the experimental procedure was safe, and later if it worked.

The bladder cancer experiment was perfect basic science — but there was no money left for basic science.

The scientists had done everything right. They had submitted all the required paperwork, as well as (importantly) several letters of support from both expert organizations like the Bladder Cancer Advocacy

---

[1] https://www.cirm.ca.gov/our-progress/awards/pluripotent-stem-cell-derived-bladder-epithelial-progenitors-definitive-cell

Network, individuals like Emilio Darpino and Vernon Brown who are struggling to stay alive with the disease, as well as myself as an interested bystander — and they had showed up. It is a lot easier to say no to a featureless name on a list, instead of an individual in the room with you, politely looking you in the eye.[2]

I had even quoted superstar John Wayne, who famously said of cancer, "I whipped the Big C!" — unfortunately he did not, and died of stomach cancer.[3]

But if there was no money? What to do, what to do?

I have a motto, which I recommend highly.

When all seems lost, keep talking.

So, we did. The scientists explained why bladder cancer research was so absolutely vital (it was) and why this particular project was excellent science (also true) — I put in my two cents' worth — and suddenly the air in the room seemed to change.

ICOC members Steve Juulsgard and Jeff Sheehy came up with an idea.

Why not shift some money — from the clinical bucket to the basic science bucket?

Senator Art Torres liked the idea, like having a friendly avalanche on your side.

After a long discussion, the committee voted YES! The idea was later taken to the full meeting of the ICOC, which accepted the recommendation.

And that is how a bladder cancer research project defied the odds, and was funded by the California stem cell program.

Two questions arise:

One: What if the scientists had not been there to fight for their project?

And two: what if there had been no funding to fight for?

---

[2] https://www.cirm.ca.gov/sites/default/files/files/agenda/Letter%20to%20the%20Board%20DISC2-11105-2.pdf

[3] https://www.cirm.ca.gov/sites/default/files/files/agenda/Letter%20to%20the%20Board%20DISC2-11105.pdf

54 Sickle Cell — And Insults?

CIRM is challenging sickle cell disease from various different angles — including Mark Walters' efforts with a gene-editing tool called CRISPR/Cas9.... (Children's Hospital of Oakland photo.)

Imagine excruciating pain, like broken glass in your bloodstream. Now imagine, on top of that, being insulted — at the hospital? Hold that thought.

First, the enemy: sickle cell disease (SCD), a change in the shape of your blood cells. Instead of being round and soft, the cells become hard and c-shaped, like a sickle, whereby the name.

These hard, sticky cells can clog the blood vessels, an agonizing experience called a "crisis." Because these crises slow the blood's ability to transport oxygen, sickle cell anemia can damage organs in your body:

liver, lungs, heart, brain; by the time an afflicted child is ten, he/she will probably have had a major stroke; by 20, there is a strong possibility of severe mental damage.

This alone is reason to take joy in the California stem cell program's funding of research like Dr. Mark Walters' efforts at the Children's Hospital of Oakland Research Institute (CHORI).

Dr. Walters hopes to use a gene-editing tool called CRISPR/Cas9 to fix what's wrong in the body's cells: remove, repair, replace.

Imagine if he was successful. In financial terms alone, the benefits would be extraordinary.

"The estimated lifetime cost of care is $9 million per person. The project aims to improve SCD therapy by preparing for a clinical trial that might cure SCD after giving back sickle gene-corrected (blood) stem cells to a person with SCD. If successful, this would be a universal life-saving and cost-saving therapy."[1]

But there is something else: a sense of wrongs, needing to be made right.

Remember those previously-mentioned insults? Imagine going to the hospital, in agony, and then when you get there, you are treated like a drug addict?

Unfortunately, this prejudice appears to be real. I interviewed four African-Americans with sickle cell knowledge; each mentioned the insults as a problem.

"Those with sickle cell see going to the hospital as going into battle," said patient advocate Nancy Rene, "They "gear up" with copies of medical records and NIH guidelines, make sure they have a diary to record inappropriate remarks from medical staff, ask a friend to come along as an advocate to help them withstand the implied racism...with which they are confronted."

---

[1] https://www.cirm.ca.gov/our-progress/awards/curing-sickle-cell-disease-crispr-cas9-genome-editing

Nancy Rene, sickle cell patient advocate. (CIRM photo[2])

Ms. Rene's grandson, now 14, had a sickle-cell caused stroke — at the age of nine months.

This experience with "implied racism," unfortunately, is not unique.

"Studies show if I as a black man and you as a white person go into the ER with severe pain, you are far more likely to be supplied with pain relief drugs than me," said Dr. Ted Love, former member of the California stem cell Board of Directors, also a Harvard/Yale scientist and CEO of a major corporation.

---

[2] https://blog.cirm.ca.gov/2016/09/29/a-patient-advocates-take-on-sickle-cell-disease-the-pain-and-the-promise/

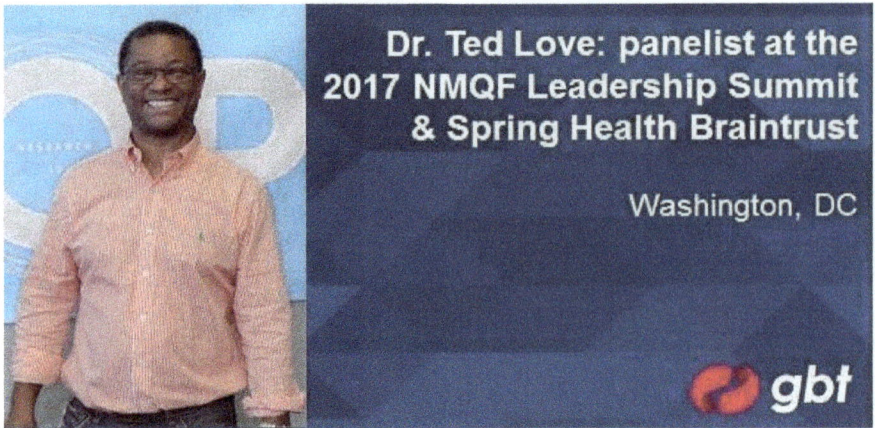

Dr. Ted Love. Global Blood Technology (Photo.)

Suggestion: California was the first state to mandate sickle cell screening at birth, mandatory now in every state. So why not have a medical card saying "SICKLE CELL PATIENT"? We already have the information, gathered by screening at birth — maybe a card could reduce the hassle at the Emergency Room?

Dr. Love had tried to retire, but fortunately, it did not stick.

I asked him how such a young man could even consider retiring, and he replied with a laugh: "Well, there is a lot of mileage on me!" But when a friend asked him to consider coming out of retirement to become the Chief Executive Officer (CEO) of Global Blood Therapies (GBT), he could not say no — because GBT is focused on fighting sickle cell.

"Ever since I was ten, kids called me Dr. Love," he said, and I could hear the smile in his voice over the phone. "I worked with patients directly for a lot of years, and it hurt to walk away from them, as if I was letting them down. But to work for a cure would benefit them more."

Now, Dr. Love's company is working on a new drug: Voxelotor.

There is also the CIRM-funded effort of Dr. Donald Kohn, famous for his work to defeat Severe Combined Immune Deficiency (SCID), which technique has cured more than four dozen children who had "bubble baby syndrome."

Dr. Donald Kohn. (UCLA Health photo.)

Perhaps a similar technique might work in the fight against sickle cell?

To remove some bone marrow from the patient, reconfigure the DNA by removing the bad mutation, and then putting the improved blood cells back — remove, repair, replace — if it works, that would be wonderful indeed.

But we cannot forget the advocate: without whom the scientist often stands alone.

When I first talked on the phone with Ade Adeyokunnu, I had a sense of recognition, as if I had known him before.

Ade Adeyokunnu, sickle cell advocate. (Linked-in photo.)

It was like a story I read about a Hell's Angel motorcycle leader, who was asked how they recruited members. He said, "We don't recruit them, we recognize them." And that is how it works for advocates. You meet someone with a disease, and they talk cheerfully and knowledgeably about it, and how it is best fought? That is an advocate, and we recognize each other.

Ade (Ah-day) Adeyokunnu is an advocate fighting sickle cell.

If you visit Sikcell.com, you can see his community website: designed so people living with sickle cell can share information with others who understand.

He should talk to Dr. Love, I thought. But would a CEO of a major corporation have time to talk with someone who might be helpful somewhere down the road?

"Of course," he said, after we had gone through the difficulties of squeezing in a half hour on his schedule to talk, and I followed up with Ade.

However — I did not hear from Ade for a while. I worried he had dropped the ball.

"No, no," said Dr. Love — "Ade is in the hospital, with an SCA crisis."

He told me a story of another person with SCA who had gone to a medical conference and then seemingly disappeared. His friends

Adrienne Shapiro, Patient Advocate of the year. (CIRM photo.)

searched for him, and finally located him in a hospital, where he had gone into a coma...

You will be relieved to know that Ade is past the crisis now. He continues to work on his Master's Degree in Business Administration — and to fight sickle cell.

I want to close with a message from one more outstanding advocate.

Adrienne Shapiro is a stem cell Ambassador for Americans for Cures Foundation. Adrienne is a great talker, very likeable, and can put into words the urgency of the need to defeat Sickle Cell.

She was just recognized as Patient Advocate of the Year by Bernard Siegel's World Stem Cell Summit.

"It's my true belief that I'm going to be the last woman in my family to have a child with sickle cell disease; and that she, Marissa's going to be the last child to suffer, and that her daughter Casey is going to be the last one to fear. Stem cells are going to fix this for us and many other families."[3]

Thank you, Adrienne: you give voice to many.

---

[3] https://www.cirm.ca.gov/our-progress/stories-hope-sickle-cell-disease-2014

55 Two Diseases, One Therapy?

Stephanie Chergui is challenging both cystinosis and Friedreich's Ataxia — with the same therapeutic weapon. (UCSD photo.)

Imagine a disease, terrible in its effects, but only a few people have it... will Big Pharma invest the billions required to find and develop its cure?

Probably not.

Lacking the possibility of profit, few corporations will risk a billion dollars plus (literally) to develop a new medicine. This is not cold-heartedness, just common sense: the biggest company can go under, if it loses too much money.

But what if there was a way to cure two rare diseases — or more — with one therapy?

Suddenly, the market multiplies.

Hold that thought.

First, consider a singularly vicious disease, Cystinosis, which stunts the growth of children. It attacks by an accumulation of tiny crystals (cystine, from which the disease takes its name) in various body organs, causing them to malfunction: in the eyes, for example, the crystals bring extreme sensitivity to light. Cystinosis also damages the liver, kidneys, muscles, brain, pancreas and more.

It is rare (only about 500 patients in America, 2,000 in the world) but deadly. It strikes before the age of two, and may lead to kidney failure before ten.

However...

In September, 2016, Dr. Stephanie Cherqui received a grant of $5.2 million from the California Institute for Regenerative Medicine (CIRM) to fight cystinosis.[1]

Dr. Cherqui has also been the recipient of major funding from the National Institute of Health (NIH) and from an amazing patient advocate group, the Cystinosis Research Foundation (CRF). Let's take a quick look at these parents in action.

CRF began with Natalie Stack, seven months old, and diagnosed with cystinosis.

I have their story from Kevin McCormack, Communications Director of CIRM:

"The family was handed a pamphlet titled, 'What to do when your child has a terminal disease'". They were told there was no cure.

The family took care of Natalie, of course, to the best of medical knowledge, but the shadow was always hanging over them. And on the eve of her 12th birthday, her mother Nancy asked Natalie what was her birthday wish.

On a napkin, Natalie wrote: "To have my disease go away forever."

The average life expectancy for people with cystinosis (then) was 18. Nancy told her husband, "We have to do something."

They launched the Cystinosis Research Foundation.

"We knew that in order to have hope, we needed research to find better treatments and a cure. In order to have research, we needed to raise funds. Without research and the hope it brings, our children would not survive."

---

[1] https://www.cirm.ca.gov/our-progress/awards/ex-vivo-transduced-autologous-human-cd34-hematopoietic-stem-cells-treatment

In their first year, the family raised $427,000, an extraordinary amount — last year they raised $4.94 million for a total of $45 million dollars — every penny of which goes to research.

How is Dr. Cherqui doing?

"We showed that (the blood stem cell transplant) in the mouse model of cystinosis...led to...long-term kidney, eye, and thyroid preservation."

On January 17th, 2019, Kevin McCormack, Communications Director for CIRM, had a wonderful announcement:

"A new clinical trial (has) just been given the go-ahead by the Food and Drug Administration..."

How does that feel to the family which has done the most in the world to battle this genuinely evil disease?

"We are thrilled that CRF's dedication to funding Dr. Cherqui's work has resulted in FDA approval for the first-ever stem cell and gene therapy treatment for individuals living with cystinosis. This approval from the FDA brings us one step closer to what we believe will be a cure for cystinosis and will be the answer to my daughter Natalie's wish made fifteen years ago, 'to have my disease go away forever.' We are so thankful to our donors and our cystinosis families who had faith and believed this day would come."

And from Dr. Cherqui?

"...the stem cells and gene therapy worked well to prevent tissue degeneration in the mouse model of cystinosis. This discovery opened new perspectives in regenerative medicine and in the treatment of a wide assortment of diseases including kidney and other genetic disorders..."

Their struggle is important not only for the fight against Cystinosis, but for many others.

As Dr. Cherqui puts it:

"If we can bring this to the finish line, we can then show the way to treat many other genetic diseases. So this could be a real benefit to a lot of suffering families."

Is there a way to get more bang to the research/therapy bullet, using similar weapons to fight different rare diseases?

Let's take a quick look at one more terrible condition — investigated by the same doctor — which might be fought with a similar weapon.

Friedreich's ataxia (FRDA) is a genetic disease, caused by a non-working body protein called Frataxin. If the Frataxin does not do its job, bad things happen.

Sufferers face life in a wheelchair. Scoliosis may warp the spine. The muscles grow weak. Heart disease, diabetes, even blindness may come — all from one condition — and as yet there is no cure.

But in late 2017, Dr. Cherqui injected blood stem cells into experimental mice, sick with a disease replicating Friedreich's.

What happened to the FRDA mice, which were given the new blood stem cells? They got better.

"Transplantation of (the stem cells)…rescued (the) impacted cells… Frataxin expression was restored. (Activity of) the brains of the…mice normalized, as did…the heart. There was also (less) muscle atrophy."[2]

The attempt to cure Friedreich's is similar to the Cystinosis effort — might we have two diseases curable by a stem cell/gene treatment: or more?

Only time, funding, and biomed can give us that answer.

P.S. Dr. Cherqui is partnering with AVROBIO, Inc. in hopes "of developing a novel and potentially transformative gene therapy for treatment of patients with cystinosis."[3]

And just days ago, January 18, 2019, Avrobio announced "… investigational gene therapy candidate for cystinosis is cleared to begin a Phase1/2 clinical trial…by the FDA…"[4]

P.S.S. Posted: June 20, 2019 Oakland, CA — Today the governing Board of the California Institute for Regenerative Medicine (CIRM) approved a grant of almost $12 million to Dr. Stephanie Cherqui at the University of California, San Diego (UCSD) to conduct a clinical trial for treatment of cystinosis.[5]

It is starting to happen. Change is on the way.

---

[2] https://www.biotechdaily.com/therapeutics/articles/294771415/stem-cell-therapy-cures-friedreichs-ataxia-in-mouse-model.html
[3] https://www.businesswire.com/news/home/20171004005177/en/AVROBIO-Expands-Rare-Disease-Pipeline-Gene-Therapy
[4] http://investors.avrobio.com/news-releases/news-release-details/avrobio-inc-announces-acceptance-investigational-new-drug-ind
[5] https://www.cirm.ca.gov/about-cirm/newsroom/press-releases/06202019/cirm-board-approves-new-clinical-trial-rare-childhood

56 Gloria at Home

What do you do, when there is no cure?

As you recall, on March 13, 2019, Gloria Reed, beloved wife of fifty years, fell and broke her shoulder in two places: the clavicle across the top of her chest, and the "ball" part of the ball and socket joint.

At first, the doctors thought she might heal on her own.

Armed with an assortment of pain medications, Gloria came home, and we waited for her to get better.

But the healing did not happen.

Two weeks later, the follow-up x-rays were ugly. The clavicle was no longer merely cracked, but broken, the bone in halves, separated by a gap.

Gloria blamed herself for the accident. "If I had only used the drier," she said.

But she had done nothing wrong: only tried to save the family money, cutting the electricity costs (one month our bill was just $50!) by hanging our wet clothes out to dry. Few people nowadays even know what a clothes pin is for, but Gloria was always so careful, with our limited family funds.

Also, regret is futile.

Remember the line from the famous poem, "THE RUBAIYAT OF OMAR KHAYAM"?

"The moving finger writes, and having writ, moves on: nor all thy piety nor wit shall lure it back to cancel half a line, nor all thy tears wash out a word of it."

But I knew what she meant, wishing we could go back and have a second chance.

Like the night Roman was paralyzed playing college football, and was lying motionless on the field. I was walking down the bleachers, step by step, thinking, this is just a mistake, surely we can go back in time, just a few seconds, and make this right...

But there is no going back. We can only go forward, the best we can.

What now? Options were bleak. To do nothing meant agony for the rest of her life, the pain masked by drugs but never gone — or, a full shoulder replacement?

"I'll take the surgery," said Gloria.

The week till surgery seemed an eternity.

But all things end except the story of the stars, and the time passed.

At last she was being operated on, while I sat in the waiting room next door, just a few feet away.

I was delighted to see Monica, Gloria's sister.

"I can only stay until 2:45," said Monica immediately.

She shared care-giving responsibility for her Mother-in-law, who had Alzheimer's disease. Monica, her husband Salvatore, and her Father-in-law divided up the care-giver chores, so the sufferer always had someone with her.

Sometimes the Mom-in-law wanted to go back to Italy, where she had been born. When this happened, the procedure was to pack her suitcase, drive around for a while and then go home, leaving her satisfied she had visited Rome once more.

When 2:45 came, Monica gave me a hug, and went back to her endless chores.

"You can go in now," said the volunteer nurse-assistant.

Then I saw her, Gloria, smiling at me, the pain lines erased from her face. I had not seen her without pain in a month.

The operation had gone smoothly, and something called a nerve block (an injection in her neck) took away the pain — but only for 18 hours. Then it returned.

After two nights in hospital (they only wanted to keep her just the one night, but even a wounded Gloria is like a wolverine in a bad mood) we drove home.

Now that she was healing (hopefully) from the operation, the drugs could be taken again. She did not much like it, being anti-druggie from way back, but it was needed, and we watched the clock to stay on schedule with the next medication.

For Gloria, who lives her life as a cheerful mad rush, existence became prison-like. From being never home, now she was restricted to quarters.

For 6 weeks she could not even drive. True, she could go with me to pick up supplies, but that meant jarring of her shoulder, which must be avoided.

When home, I can be at Gloria's beck and call. She is nice about it, always saying please and thank you — but every second there is another meal to be prepared or a bed to be made or dishes to be washed, or laundry to be done.

And of course there was the personal care: showers to be given, clothes to be helped with, drinks of water to be brought.

One question which might have been scary: how would we maintain our income, if I could not come to work?

Fortunately, I have a great boss, a genuinely kind man, and there was no question a way would be worked out so I could work from home for a while, if it was needed. I had seen him do it before.

But what about people with an uncaring boss — what happens to them? This is yet another reason America needs a standardized medical care program, so accidents like Gloria's will not be financially as well as physically crippling.

For us, that particular problem will not even come up, as Gloria insists that I will maintain my normal work schedule, and drive across the Bay three days a week: Tuesday, Thursday and Friday, to my little stem cell office.

"You must still keep a life of your own," she said.

This is genuinely noble of her. She is essentially one-handed until the shoulder accepts the foreign object, the plastic ball and socket joint now permanently in her shoulder. So she is home right now, while I am at work doing this. I have called her twice, and feel guilty not being at her side.

Kaiser sent out a Physical Therapist, who gave us hints on how to get into and out of the couch and the bed, and gave critiques on Gloria's exercise.

"You must exercise a lot," she says, "Carefully and gently, but often — you do not want scarring to freeze the joint."

Our daughter Desiree has a frantically busy schedule as the Athletic Director for the University of Nevada at Las Vegas. But she took three days off to come and be with her Mom, and her visit was a delight.

Roman brought his children over to visit Grandma, and Katherine and Jason were great medicine.

But then Desiree got on the airplane, and returned to her normal existence.

Being the cheerful outgoing person she is, Gloria has friends, and they are helping immeasurably. JoAnn sent lasagnas, Jan the flower arranger spent half a day with her, as did Carol ("Tell me what to do, I did not come just to sit and stare at you!") and other friends from her Church and social circle. This was more than food — it was the bringing of the outside world to her.

Sunday we had an outing, a cautious quarter-mile walk to church. I walked on her left side, her good arm in mine. She smelled the fresh Spring day, and enjoyed the motion.

I was dreading the removal of her shoulder bandages. The massive taping somehow was a comfort, as if the clean white strips held her together.

"Take off the outside bandage after four days," the doctor had said, "But leave the smaller bandage underneath."

But we did it, carefully, with Gloria in charge every step in the way, and the tape was gone. She looked younger, stronger, with her normal pink skin showing at the neck and shoulder, and that huge bandage gone.

In six weeks, or so, Gloria should be better. She will be physically diminished, probably never again able to raise her arm above shoulder height, and her life will be challenged. But hopefully, the pain will grow less.

I think of Roman, paralyzed 24 years, but still defiant: still pushing the next incarnation of the research funding bill named after him: AB 214 (Mullin) was unfortunately just shot down in committee. But we will try again next year, and sooner or later they will all get tired of telling us no.

What a shining miracle it would be, if both Mother and son could be healed!

If cure does not come in time to help our families, then others like ours, equally deserving of cure, sooner or later, will gain relief from suffering.

To become well...

It feels impossible, of course. There is no path to instant cure; we must endure the slow and careful way, like stepping from rock to rock across a rushing stream.

With Roman, it seems he has been paralyzed forever. His skin breaks down now, which it never did before, and he gets pressure sores and all the other ills paralysis brings. But yet he will fight on, and so must we.

And this is why the California stem cell agency exists: to seek relief for "incurable" injuries and conditions: to heal bones faster and completely, to end paralysis and cancer, and heart disease and stroke, and so many other conditions which afflict the ones we love.

Where there is an incurable disease, CIRM will fight it. Where there is a breakdown in progress, CIRM will send scientists to take it apart, and fix it.

And if you read their paths to progress, you may perhaps come to share my view:

If the research can continue, cure is not just possible, but inevitable.

57 Of CIRM, and Buying My House

The building where CIRM lives: Oakland, California. (Stem Cell Report photo.)

In 1997, my wife Gloria and I bought a house. The asking price was somewhat above our comfort zone ($165,000, a lot back then) but we bought it anyway, going deep in debt. Just three days later, we had a visit from the same realtor who sold us the house. He wanted to buy it back — for $215,00 — $50,000 more than we had paid. We said no, we liked the house (and live there still) but it was a pleasant surprise to find unexpected value in our home.

Like our house, the most sensible investment the Reed family ever made, the California stem cell program is packed with benefits, many of a financial nature.

As we move closer to decision-making time, when California will vote whether or not to renew CIRM's funding, we need to count the cost. The opposition will almost certainly try to make CIRM seem a wild-eyed foolish expenditure, so it makes sense to consider such things beforehand.

CIRM's original cost of $3 billion dollars (plus $3 billion interest) is being paid back over a 30-year time span, not unlike buying a house. It is expensive (though nothing compared to the mountainous and ever-increasing costs of disease) so is being paid back by a sensible long-term method.

Is backing stem cell research a prudent investment?

First, California is operating at what the Los Angeles Times calls "a huge surplus" (more than $21 billion) right now.[1]

When you consider CIRM cost "only" $300 million a year, one-third of a billion, it seems to me we can definitely afford it.[2]

Yet money should never be spent without figuring the economic impact.

So: if you go to page 1 of the Prop 71 Economic Impact Analysis conducted by Stanford University Analysis Group's Professor Lawrence Baker, you will find a paragraph jam-packed with meaning[3]:

"The…impact of the Institute is increased by three factors:

1.   …donor matching funds
2.   Research leverage as researchers…obtain grants from other sources

---

[1] https://www.latimes.com/opinion/enterthefray/la-ol-newsom-budget-flush-taxes-20190509-story.html
[2] https://www.the-scientist.com/news-opinion/stem-cell-funding-agency-cirm-is-nearly-out-of-funds-66108
[3] http://etopiamedia.net/empnn/pdfs/analysisgroup1.pdf

3. The economic multiplier effects as the funding moves through various levels of the economy..."

What is the "multiplier effect"?

"The multiplier theory (was) created by British economist John Maynard Keynes. Keynes believed that any injection of government spending created a proportional increase in overall income for the population, since the extra spending would carry through the economy."[4]

We mentioned earlier that CIRM has attracted $3.1 billion in new money already. (More recent estimates run higher.) Not to mention, many of the new businesses (created by research funded by CIRM) will likely continue to contribute to the state's economy.

It is reasonable to expect the renewal of CIRM to do the same.

"And of course all that additional money creates additional jobs...and helps further develop the industry here and further cements California's position as a global leader in the field." — Kevin McCormack, personal communication.

The closer you look at CIRM, the more benefits you find.

Take another look back to the Stanford University analysis group:

"We focused on 6 of the 70 conditions identified by medical and scientific experts as having the potential to benefit from stem cell research. These include(d) stroke, heart attack, diabetes, Parkinson's, spinal cord injury, and Alzheimer's...

"In our models, a small cost reduction of 1% per year...for these 6 conditions...produces cost savings to all Californians of $11 billion over the 35 years analyzed."[5]

http://www.etopiamedia.net/empnn/pdfs/deal-baker1.pdf

To my mind, a one per cent cost reduction is a minimal projection. For instance, what does it mean to a quadriplegic (previously paralyzed in both upper and lower body) to regain the use of hands and arms? He/she may now be able to eat on their own, or require less personal maintenance. Perhaps he/she will no longer require 40 hours-a-week attendant care — instead getting by on far less. This is concrete dollar-and-cents finance, when hiring an attendant may cost $32–$40 an hour...

---

[4] https://www.investopedia.com/terms/m/multiplier.asp
[5] http://www.etopiamedia.net/empnn/pdfs/deal-baker1.pdf

More benefits: Cure science depends on mice and lab rats. But where do you get them? These are not just rodents you catch in somebody's garage! You may want mice with a certain disease, or, lacking an immune system, or other characteristics.

CIRM needed medical mice. They found a company which specialized in that — but it was in Maine — and we can only pay to California. What happened?

"The Jackson Laboratory in Maine...greatly expanded their West Coast operation in 2009, when they received CIRM funding... establishing an independent West Coast operation that today employs approximately 120 people."[6]

See how that worked? There was a medical need, which CIRM answered. An out-of-state firm relocated here in California — 120 new jobs were created in the Golden State — and a lot of families benefited.

California is where the biomed action is, and CIRM is a major reason why.

---

[6] https://www.cirm.ca.gov/about-cirm/2011-report-therapy-economy

58 Cooperation with the Capitol

Every time I see our State Capitol, I feel such pride! (Photo by Wikipedia.com.)

I wish you could have been there, August 15, 2018, at the state Capitol building, Sacramento, California. It was a Biomed Committee hearing for the California Institute for Regenerative Medicine (CIRM), formed by the citizens' initiative, Proposition 71, the Stem Cells for Research and Cures Act.

The parking lots close to the Capitol were full, but at last I found an outdoor lot, negotiated my ten dollar bill into the right machine, put the receipt on my dashboard so I would not be towed, and took off walking.

As always, it was a joy to see the white-domed Capitol building, symbol of our state's democracy.[1]

Senator Art Torres is very much at home in the halls of Sacramento. (CIRM photo.)

On the green lawn outside were hundreds of people, full of energy, excited to be here whatever their reasons: to fight an injustice, bring peaceful improvement, or maybe preserve something wonderful — like CIRM, California's Stem Cell Agency.

Inside, glass-walled display cases showed off the history and products of our state; in front of the Governor's office was a giant statue of the California state animal, the grizzly bear, gift of former Governor Arnold Schwarzenegger. And above? Ceilings so high you could fly a modest airplane, or dream of a better world.

A short ride on an oak-paneled elevator brought me to the 4th floor, and a hallway between two buildings, the old and the new portions of the Capitol.

Room 444 was where the hearing would take place, but the door was locked. The waiting area was like a painting: a wooden lattice framed blue sky, while soft-pile carpet absorbed our footsteps.

---

[1] Photo by Andre m — Own work, CC BY-SA 3.0, https://commons.wikimedia. org/w/index.php?curid=31832691

I was two hours early, partly because I can't stand being late, but mainly because I would be allowed to make a public comment, five minutes to talk about CIRM. What an opportunity! I have no official connection with the California stem cell program; but it is a citizen's initiative, brought into being by the people of California.

But now CIRM's funding was running low; it was vital that it be renewed, to continue its great work.

Today was a time to remind Sacramento legislators what CIRM has done with the trust, and the money, that California placed in it.

I sat down on a deep-cushioned chair, intending to read my notes... but the chair was sooooo comfortable...

My eyes opened when CIRM folk invaded the premises: leadership people, whom I would point out to you, if you were new at a public meeting, and sat next to me.

Here was Senator Art Torres (Ret.) with that spectacular mane of white hair. Vice Chair of the board of directors, Art could not relax for long, because every few seconds another Senator or Representative would approach, notice Art, and stop to shake his hand. I remember once he and I walked into a committee meeting, and the chair stopped the meeting just to come over, saying his father had once worked with Senator Art.

CIRM President and CEO Maria Millan joined us: a cheerful powerhouse, beaming energy. You knew immediately she was the President; very strong, someone you would want beside you when it was time to fight the dragons, but still approachable.

Maria Bonneville bounded into the area. Carrying herself like a runner, she needs lots of energy as V.P. of Administration: I think of her most as the link to patient advocates.

Chila Silva-Martin was there; I asked was she going to speak, and she said, no, not unless someone had a question about numbers — Ms. Silva-Martin has the aching weight of responsibility, to understand and explain CIRM's financial plan, where every nickel goes. I would not want her job, but I am glad she is doing it.

Chila Silva-Martin, finance officer. (CIRM photo.)

Scott Tocher rushed by, nod and a wave and on his way. As general counsel, he was there for legal stuff; to know every policy that makes the program work, and make sure they were being complied with.

Scott Tocher, CIRM's legal eagle. (CIRM photo.)

If they gave an award for Most Cheerful CIRM employee, that would be a shoo-in: Kevin McCormack, communications director. As usual he was there chatting with folks; Kevin begins by assuming you are a friend, and isn't it great we have this wonderful program — let's work together, it will be fun!

Kevin McCormack, serious individual! (CIRM photo.)

Conspicuous by his absence was Bob Klein, who technically does not work for CIRM anymore, but is the heart and soul of it nonetheless. The man who began Prop 71 was hosting a birthday party (his own!) on a boat in San Francisco Bay.

"They're opening the doors!", someone said.

I sat down at the front table, next to President Millan and Senator Torres.

This was the busiest time of year for legislators, and many committee members would be "on the floor" in the main voting chambers, making sure the bills they supported or opposed were dealt with.

Chairman Kevin Mullin (D-South San Francisco) opened the meeting. He has a deep commitment to biomed, especially as his district has been called "the largest biotechnology cluster in the world" by the California Life Sciences Association.

Assemblymember Kevin Mullin, biomed champion. (Official photo.)

Assemblymember Bill Quirk (20th District) was a veteran of many legislative battles in support of biomed; he asked numerous questions, said he was "satisfied" with CIRM's delivery.

Assemblymember Bill Quirk, pioneering biotech supporter. (Official photo.)

Assemblymember Marie Waldron of the 75th District, Minority Floor Leader, was very positive; I found out later she had been honored as "Legislator of the Year" by the California Chronic Care Association — https://ad75.asmrc.org/#biography

Assemblymember Marie Waldron, award-winning legislator. (Official photo.)

Assemblymember Todd Gloria was supportive, but new to me; I found out later he had been elected in 2016. His bio held something I liked: "a lesson his parents — a hotel maid and a gardener — taught him at a young age: if you truly care about something, then you should leave it better than you found it." — words to live by.

Assemblymember Todd Gloria's parents inspired him to make a better world. (Official photo.)

CIRM President Maria Millan had the task of giving an overview of CIRM. This was clearly impossible, but she did it anyway. Here is a typical paragraph:

"CIRM has funded over 800 projects at over 70 institutions and is the largest single funder of clinical research for stem cell regenerative medicine. In 2016, we launched a bold 5-year strategic plan to cut the time in half for the development of a regenerative medicine therapeutic and to markedly increase the number of clinical trials. To date, CIRM has funded the treatment of over 900 patients in these clinical trials for a broad range of indications including fatal pediatric orphan disease, stroke, ALS, spinal cord injury, heart disease, chemotherapy resistant advanced cancer, diabetes, blinding eye disease, and sickle cell disease."

She said A LOT in a relatively short period of time. To hear her speech, go to[2]:

Senator Art Torres gave brief remarks, centering on two of his "favorite CIRM programs," SPARK, the Summer Program to Advance Regenerative medicine Knowledge, and the Bridges to Stem Cell Research program, both providing funds for deserving but low-income students, building their involvement in a stem cell career.[3]

Jan Nolta, champion of hard work. (UC Davis photo.)

Jan Nolta of the UC Davis Institute for Regenerative Cures gave a crisp presentation, citing the success of her institute as "all because of

---

[2] https://www.assembly.ca.gov/media/assembly-select-committee-biotechnology-20180815/video

[3] https://blog.cirm.ca.gov/category/events/cirm-bridges-program/

CIRM," for funding the scientists. She had a funny word for the terrible disaster it would be to not have CIRM funding renewed, calling it "CIRMageddon" instead of Armageddon; she also provided a crucial statistic about the National Institutes of Health (NIH), where "only 5–8% of proposed grants get funded."

David Jensen, editor of the massive weblog, California Stem Cell Report, gave a strong presentation. The "king of the critics" of CIRM, he may be counted on to point out his disagreements with the program — but he is a reporter, and will present positives as well as negatives. (His report began with a statement about Mazatlan, where he once lived on a boat, so I used that picture.)

For example: "The agency operates with financial and oversight autonomy that is rare in California's government, courtesy of the ballot initiative that created it. But that measure also proved to be both a blessing and a curse. The agency's financial autonomy has allowed it to provide a reasonably steady stream of cash over a number of years, something that is necessary to sustain the long-term research that is critical for development of widely available treatments. At the same time, the ballot measure carried the agency's death warrant — no more money after the $3 billion is gone. Cash for new awards is now expected to run out at the end of next year..."

A transcript of his remarks is available at[4]:

Jensen in Mazatlan (personal photo).

---

[4] https://blog.cirm.ca.gov/2018/08/21/a-brief-history-of-the-stem-cell-agency/

Senator Torres introduced a family which benefited from CIRM's investment:

"Mr. Pawash Kashyap and Upsana Takur are the loving parents of (baby) Ronnie Kashyap who suffered from X-linked SCID (Severe Combined Immune Disorder), a life-threatening immune disorder that left him unable to fight infections. UCSF found a way to cure Ronnie... now, because of CIRM-funded research, he is a beautiful little boy with no threat of life-threatening infections..."

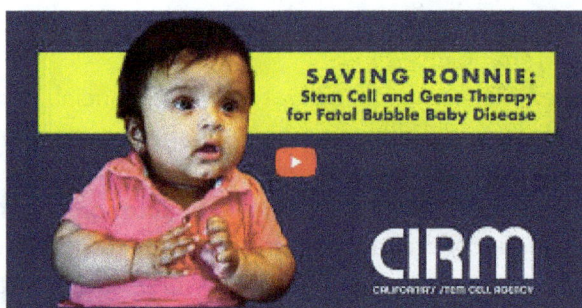

What could be more important, than to save a child's life?

My favorite moment: when President Millan and Senator Torres added up the amount spent ($2.78 billion) and compared it to the money brought in by add-on grants and other sources ($2.7 billion dollars), showing we are getting more bang for the buck.[5]

As a patient advocate, I have a lot to say, and it was difficult to reduce it to just five minutes — but I managed to restrain myself somehow, and the speech is below.

"My name is Don Reed. I am the father of a paralyzed young man, Roman Reed, and the author of two books on the California stem cell program.

If you have a chronic disease, or provide care for someone who does, you know what it is like to have been told — the condition is incurable; there is no hope.

When Bob Klein inspired and led the citizens' initiative, Proposition 71 in 2004, we did have hope, like never before: that stem cell research might alleviate, or cure many chronic ills afflicting our loved ones.

---

[5] https://www.cirm.ca.gov/our-progress/funded-institutions

Today, thanks to the 7.2 million voters who authorized the California Institute for Regenerative Medicine, or CIRM, we have something better than hope; we have results, accomplishments, people made well — and a systematic way to fight chronic disease.

How does it work? Scientists apply for grants, a board of experts from outside the state evaluates the proposals, in-house scientists give their opinions, and the Independent Citizens Oversight Committee — with the help of the public — makes the final decision.

Curing disease is not fast, nothing guaranteed; there will be setbacks. But the CIRM way is like a farmer planting seeds. Results do not come overnight. But if the preparations are done right, cures — like crops — become almost inevitable.

It is happening now. Therapies are advancing, patients are benefiting, knowledge is being shared. Over 2,700 medical discoveries have been peer-reviewed and published in scientific journals — pieces of the puzzle of cure.

Children's lives have been saved: fifty beautiful little munchkins who can now run around outside, happy in the sunshine, cured, actually cured, of the Bubble Baby Disease. Before, they had no immune system — now they do.

A young man, Kris Boesen, formerly completely paralyzed, is now lifting weights. He received a stem cell therapy originally developed by the Roman Reed Spinal Cord Injury Research Act, and advanced by CIRM. Kris Boesen recovered the use of his arms and hands. Think what that means to a paralyzed person.

Brenden Whittaker had granulomatous disease, a vicious immune disorder that prevents him fighting off fungal or bacterial infections. Over the years multiple infections had eaten away at his lungs and liver. His early years involved literally hundreds of trips to the hospital; his chances of surviving to adulthood were slim. He had a looming death sentence. But today, thanks to therapies developed with the help of CIRM, Brenden Whittaker lives. Wanting to pay back CIRM for his life — he decided to become a doctor, now that he will have enough years for that career.

CIRM is part of my favorite medical research story: in Toronto, Canada, three scientists entered a big room full of young people in beds, with their families around them. The children were in comas, unresponsive: dying. But the scientists had a new substance to try, made from the pancreases of dogs. With the parents' permission, they started at one end of the room, injecting the comatose patients. When they reached the far end, the children who got the shots first — were sitting up and talking to their astonished parents. It was *insulin,* and it saved their lives. But even so — all those shots — this was a temporary fix, not a permanent solution.

In 1977, California developed artificial insulin — and today, CIRM is building on that legacy — to develop something like an artificial pancreas, to be permanently implanted in the body, where it will convert stem cells into insulin.

As for Roman, my son, cure has not come yet. He remains paralyzed with a chronic injury, and I am his care-giver. At 73, my health is good. But what will happen to him, when I die? I ache for him to be well, to do all the things able-bodied folk take for granted. No one should suffer what he goes through every day — and California has three million citizens with a disability.[6]

Maybe, in years to come, Washington will become more responsive to the medical needs of all our people, including the 50% of our population living with at least one chronic disease, like heart disease, cancer, stroke or diabetes. But hope is not enough. We cannot sit idly by and just wait.[7]

We need something proven, practical, and here right now: CIRM, the glory of a state, the pride of our nation, a friend to all the world.

Please do everything in your power — to support, protect, <u>and renew</u> — the California Institute for Regenerative Medicine. Thank you."

P.S. If you want to see the hearing for yourself, here is a link to a video, provided by Jason Stewart of Americans for Cures Foundation.[8]

---

[6] https://tinyurl.com/yawlejay

[7] https://www.cdc.gov/chronicdisease/

[8] https://www.assembly.ca.gov/media/assembly-select-committee-biotechnology-20180815/video

59 The Christmas Truce

Imagine ending a World War — for one day.... (Photo by Wikipedia.com.)

In the nightmare battlefields of World War One, three things stand out to me.

First, the ugliness of trench warfare, where men dug ditches and lived in them, and, every so often, climbed out of the trenches and screamingly advanced toward the next set of ditches, in which other men were living, and each side shot and killed each other as best they could. Those who made it to the next trench lived and the others died in the mud, in places with names like Verdun and the Somme.

Second was the mass use of gas warfare, clouds of drifting yellow smoke which slaughtered indiscriminately, killing and killing, and when the winds shifted, the death-air might blow back on those who had sent it, or innocent villages nearby.

I breathed it during basic training in the Army, 1963. Ten of us were given ugly black rubber masks, and told we would be experiencing poison gas.

"There are two kinds," the sergeant said, "CN and CS — we call them Cry Now, and Cry Sooner — that is the worst — guess which one you're getting?"

We would go into a small airtight room, and when the door was closed behind us, poison gas would be released. We had to take off our gas masks, and breathe poison until we all had at least one good gasp of it — then we would put our masks back on. The idea was to trust both our leadership and our equipment.

In the gas room was a bench, and we all sat down on it.

Note: to see this in action, go to a military website[1]:

The masks we wore made us look like monsters. Inside them we could hear the roar of our own breathing, like the scuba-tank rrrrrrr-haaaaa of Darth Vader in STAR WARS. Sweat emerged where the rubber clamped onto my face.

HSSSSSSSS A grey cloud entered the room.

We looked at the Sergeant, who held up his hand.

HSSSSSSS — the cloud increased until we could barely see each other.

"Okay, boys, take 'em off," said the Sergeant.

I breathed in Hell.

It felt like burning fingers thrust up my nose. The skin on my eyeballs dried, like hardboiled eggs.

"Take a breath," the Sergeant said.

I was coughing, choking, trying not to vomit — "Wait a minute," said the Sergeant, "Smith does not have his gas mask off."

Through the cloud of gas I saw Smith, who still had his mask on. He must have heard us, because he clutched his hands to his gas mask — to take it off?

No. To keep it on.

"Help him," the Sergeant said.

We jumped on poor Smith, and encouraged him. At last, the mask was off. And yet?

---

[1] https://www.youtube.com/watch?v=guD79qdvy1Q

"He's holding his breath," the Sergeant said, "Everybody has to take a breath."

It seemed eternity but at last Smith choked and coughed, and the Sergeant said okay and we all put our gas masks back on. The hissing stopped and a door opened up and we tumbled out into the clean air world.

Smith threw up and everybody hated him for a while.

But for those who breathed the gas clouds of Verdun, there was no clean air escape; and those who survived the gas warfare in the trenches? They took the damage home in their lungs, to suffer the rest of their lives.

But there was also one day, Christmas Eve, 1918.

No one remembers which side came out first, waving a white flag — not surrendering, nothing like that — just, wait a minute, cease firing.

The other side came out too, slowly, clambering awkwardly up the walls of mud.

One brought his gun, but motions were made and he handed it back down.

The men in the armies of both sides got out of their trenches, approached each other, and shook hands.

Alcohol was shared, and clumsy attempts at language — "Sprechen Sie Deutsch?" "Ja, ein wenig, aber nicht so gut!" — and for an hour (in some places 2 days!), peace reigned.

And then in the distance a flare went off. Both sides looked at each other, and sighed, and shook hands again, and returned to their trenches, and set about killing each other once more.

That sweet moment became known as the Christmas Truce. The Generals hated it, but I am not a General, and I love that brief instant of sensibility.[2]

America is divided today, as it has not been for many years.

Republicans control the Senate and the White House, and appear to think their President is wonderful.

Democrats control the House, but are otherwise excluded from power, and have a different opinion of the man in the White House. The cities are Democrats, suburbs lean that way, and the rural parts are

---

[2] https://www.history.com/topics/world-war-i/christmas-truce-of-1914

Republican. The West Coast is one way, the South is another, the Mid West is tugged in two directions.

Every night I watch the news and shake my head — sometimes my fist — at the seemingly endless political war.

Sometimes it seems nothing good can come, unless one side overwhelms the other — naturally I want the winning side to be mine — as I'm sure the other side does too.

A recipe for chaos? Or gridlock?

And yet, in the middle of all this, a moment of truce.

Democrats and Republicans united to pass the Criminal Justice Reform Bill — with the support of the President. From Ted Cruz of Texas to Kamala Harris of California, ideological opposites found common ground.

Naturally, not everyone was satisfied. Senator Harris called it a "compromise of a compromise", and twelve conservative Republicans voted against it, as "soft on crime."

Even so, on a vote of 87–12, the act passed the Senate. The House did the same — and the President signed it into law.

Good job, Government!

Let's do that some more, working together, beyond politics — like maybe fighting for a cure for cancer?

Here are two political leaders, opponents: but they both want the same thing.

"President Trump made a new promise if voters grant him a second term: "We will come up with the cures to many, many problems, to many, many diseases, including cancer."

"Trump's statement...echoed remarks by former vice president and Democratic candidate Joe Biden on the stump last week in Iowa: "I promise you, if I'm elected president, you're going to see the single most important thing that changes America: We're going to cure cancer."[3]

Seems like another opportunity for a Christmas truce.

---

[3] https://www.realclearscience.com/2019/06/24/039curing039_cancer_easy_politics_near_impossible_science_286440.html

60 Alexander's Challenge

Roxana, the most beautiful woman in the world — in an unconquerable city — and Alexander the Great outside the walls.... (Pinterest photo.)

When Alexander the Great had defeated Darius III, and was tromping through Persia, he heard about a queen named Roxanne, princess of Bactria, said to be the most beautiful woman in the world.

Being Alexander, he could do no less than to go find out, and so he and his weary troops walked miles and miles until they came a high-walled city.

At the top of the wall stood a woman, and she called out to Alexander, "Who are you, and what do you want?" (As if she had not already known who he was for his past hundred miles of travel!)

And he replied, "I am Alexander of Macedon — and I intend to marry you!"

The soldiers on the wall laughed. The soldiers on the ground did not. They knew Alexander.

And Roxanne said: "How can you hope to conquer my city — when your troops are so tired and weak and hungry they can barely stand up? Let me feed them!"

She raised one arm, and the catapults fired, hurling objects through the air — loaves of bread, heads of lettuce, bunches of grapes.

It was a serious gesture. To conquer a city, besiegers use patience, waiting outside while the inhabitants starve — yet here was a city so confident of its food supply that it threw large quantities away...

Alexander thanked her. The men picked up the food. They walked away, retreating to the banks of a nearby river, where they bathed, drank, rested — and brooded how they could break the unconquerable city.

But their home country was Macedon, by the sea, and the people then were great breath-holding divers, and presently one of them came out of the river and ran to Alexander's tent. That night, Alexander and the best swimmers went into the river again, where presently they ducked underwater and disappeared.

Using wineskins for a short-term air supply, they swam through the underwater opening the diver had found — and came up in a great well inside the city.

The inhabitants of Roxanne's city were shocked at the sudden emergence of Alexander's soldiers, climbing out of their water source — the fight was short-lived.

Alexander stayed with Roxanne for several years, not only marrying her, but presiding over the mass marriage of hundreds of his own men and the city's women.

I love the part where Alexander shouts, "I will marry you!" — but there are quieter kinds of challenges as well...[1]

There is a reason I chose the title of my book: "REVOLUTIONARY THERAPIES: How the California Stem Cell Agency Saved Lives, Eased Suffering, and Changed the Face of Medicine Forever."

---

[1] https://en.wikipedia.org/wiki/Roxana

For one thing, it is true: The California stem cell agency has indeed saved lives, eased suffering, and changed the face of medicine.

But the first two words, "REVOLUTIONARY THERAPIES," were suggested by the man who began California's stem cell program.

If you know stem cells, you know Bob Klein, the driving force behind CIRM. Many helped, but he is the one irreplaceable man.

And now the money behind CIRM's funding has almost run out.

The citizen's initiative must rise again, if we are to renew the program.

It is so important — the National Health Council estimates that "chronic diseases affect approximately 133 million Americans...(and) by 2020 that number is expected to grow to an estimated 157 million, with 81 million having multiple conditions..."[2]

The cost of chronic disease? Nearly $3 trillion dollars...[3]

But will Klein lead the fight again, taking on the tremendous effort once more?

In a way, he already has. The group he leads, the small but dedicated staff of Americans for Cures Foundation are already doing behind the scenes chores.

Top-notch people, including:

Melissa King, hard-working veteran of the first campaign.

Jason Stewart, cheerful, politically savvy, knows everyone.

Mitra Hooshmand, Ph.D., very knowledgeable on regenerative medicine.

Robert Klein III, son of Bob, very astute in governmental programs.

Anna Maybach, our newest member, analyzes stem cell and gene therapy programs.

And me you know already.

We share one goal: to let people know what an incredible success the California stem cell program has already been, and why we must fund it again.

Everybody wants Bob to take on the $5 billion renewal, but it is easy to wish a mountain of work on somebody else. Bob personally contributed roughly $3.4 million to the first campaign, borrowing to raise it, and that was only the financial side; bigger was the ceaseless chores of total involvement, to turn a dream into reality.

---

[2] http://www.nationalhealthcouncil.org/newsroom/about-chronic-conditions
[3] https://www.cdc.gov/chronicdisease/about/costs/index.htm

Four leaders of Americans for Cures Foundation, the fight to protect and renew the California Institute for Regenerative Medicine. From left to right, they are: Jason Stewart, Advocacy Coordinator; Melissa King, Field Coordinator; Mitra Hooshmand, Science Director; and Robert Klein III, Director of Government Relations. (Americans for Cures Foundation photos.)

Welcome to Anna Maybach, newest member of the team, our Stem Cell and Genetic Research and Therapies Program Analyst.

Now Bob has done so much in his life (and he does not get to read or edit this) you might think he talks about himself a lot. But that would be wrong.

Bob is mission-driven. Almost every word that comes out of his mouth is to:

A. ask about you and your family, or
B. advance the cause.

He works in the real estate field, developing housing complexes, all of which contain a substantial proportion of low-income homes. It is my belief he is trying to save up money to finance the second campaign.

Knowing the demands on his time, I carefully plan my words to him, as few as possible. When I need 30 seconds, I ask Elizabeth Tafeen, Bob's all-knowing personal assistant, and we strategize, like maybe tomorrow at 1:46.

Of course, if he drops by my little office and slides back the door to say good morning, well, he is fair game then!

And it is not impossible that I might accidentally wait outside the restroom, to say, "Oh, hi, Bob, how are you doing? Listen, what do you think about...?"

But it was in his office, I think, that he made up his mind, about the second instalment of CIRM. It is a beautiful office. A giant desk, all wood paneling. Windows looking out over trees. Pictures of young Bob with young Bill Clinton, lots of his family, and a bunch of other celebrities, mostly political.

Now one of my greatest joys is to come across something terrific about CIRM, and share it with Bob. You might think he knows everything that goes on at the institution he built; but no, CIRM is a different world, and he has to let go of it to do the second half of his life, which makes possible the first.

On this day, Dr. Jill Helms had gotten an almost six and a half-million dollar grant from CIRM. This would allow her to continue her work on bone reconstruction, to help the 6.8 million Americans who suffer broken bones every year.[4]

---

[4] https://www.cirm.ca.gov/our-progress/people/jill-helms-0

And I brought a page about it to Bob, and said, "No action required; this is just something wonderful CIRM made possible."

He glanced at it and nodded. And then he said:

"The scientists are worried they won't have the funding."

And then, very softly, as if he was alone in the room:

"But I will get it for them."

61 The Smallest Miracle

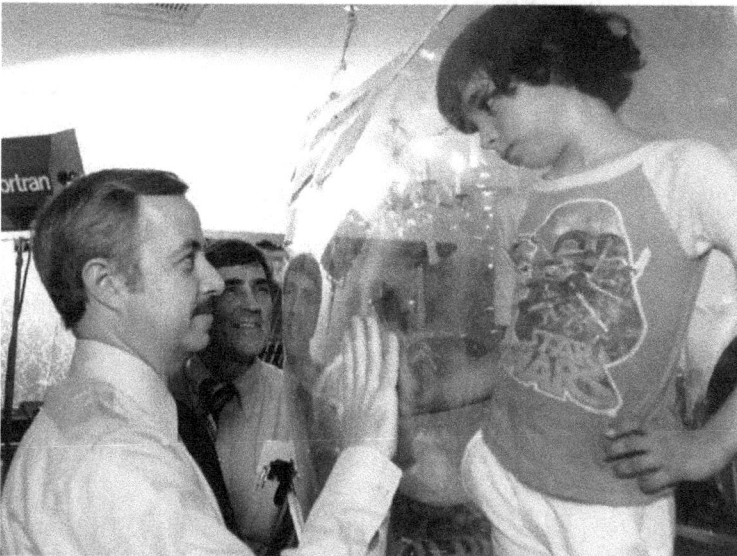

Dr. William Shearer meets David Vetter, the boy in the plastic bubble. (Photo by Texas Children's Hospital.)

Did you see a movie called "THE BOY IN THE PLASTIC BUBBLE," starring John Travolta? In the film, the hero has a disease called Severe Combined Immune Disorder (SCID), which means his immune system does not work. Germs that you and I would not even know about (because our immune system would fight them off) could be fatal. To survive, he must live in sterile housing: the plastic bubble.

The movie was based on the true story of a boy named David Vetter.

In the Travolta movie, the boy in the bubble grows up, falls in love, and decides to leave his plastic refuge. His movie doctor tells him he

Evangelina Vaccaro, safe and well, introduces herself to the CIRM board, saying "Thank you." (CIRM photo.)

may have built up some immunities; he steps out into the world, and rides off on a horse, behind his girl.

The real-life story did not have such a happy ending.

David Vetter's family fought for him with tenacity, intelligence, and love. They managed to get him tremendous amounts of treatment, including help from the National Aeronautics and Space Administration (NASA), which built him a space-suit device to wear, so he could go for a walk outside. It was cumbersome, and he only used it seven times.

The bubble had been set up as a temporary device. The family had lost another child to SCID, and this time they hoped science might provide them a cure, if given time. But when cure did not arrive, the boy stayed in the bubble, to survive.

When his doctor, William T. Shearer, introduced himself, David Vetter put his hand up to the plastic wall — and on the other side, Dr. Shearer put his hand up to meet it.

https://primaryimmune.org/living-pi-explaining-pi-others/story-david

When David reached 12, a transplant procedure was tried. His sister Katherine volunteered to give cells to David in a bone marrow transplant, just as she had done for his brother, years ago. At first, it seemed to be working. But hidden in the young woman's body was traces of a dormant virus, Epstein–Barr, undetectable in the screening. It triggered a cancer, which overwhelmed the boy's body.

At last David said to his doctor, "Here we have all of these tubes and all of these tests, and nothing's working. I'm getting tired. Why don't we pull all of these tubes out and let me go home?"

When David could no longer be effectively treated in the bubble, he was taken out: February 7, 1984. His mother kissed him for the first time. But the lymphoma had spread throughout his body. His health deteriorated.

Importantly, David knew every step of the way what was going on, to the limits of his understanding. When he was four, he took a syringe and poked a hole in the bubble. Germs were explained to him, and from that point on this very bright child was kept fully informed.

His last conscious act was to wink at the doctor who had tried so hard to help him.

Then he passed away.

And this was the condition which threatened baby Evangelina Padilla Vaccaro.

"I knew something was wrong right away," said her mother Alysia in a telephone interview:

"She had a gray color, did not move right, and she spit up a lot... she had no immune system; her body had no way to fight back against germs; a cold could kill her. We made our house as sterile as we could. My husband Christian and I wore masks all the time. Evangelina never saw our mouths."

"And then I heard about stem cells and UCLA and Dr. Donald Kohn — and how it might be possible to rebuild Evangelina's immune system.

"But stem cells? I had been raised a good Catholic girl, and had heard all kinds of bad things about stem cells. But when it is your child in danger, you will listen very carefully, and we did. This was about saving lives.

"There was just one opening left on the clinical trials...Dr. Kohn held it open for the two months it took our daughter to grow strong enough to endure the operation.

"Dr. Kohn was up front with us...never promising a cure. But my husband and I had read more than 30 studies on the National Institutes of Health website, trying to educate ourselves on what had been tried before on our daughter's condition, and Dr. Kohn's approach made sense.

Dr. Donald Kohn, NEWSROOM.UCLA.EDU.

"He would take out bone marrow from her hip. There was a mutation in her genes. He would fix the gene, put it into some stem cells, and put those back.

"I had had three miscarriages before the twins arrived...was I now to lose another of my children?"

And then, everything changed. Her mom and dad took Evangelina outside, and held her high in the air. She felt the sunlight for the first time. She opened her mouth, and laughed for sheer joy.

Evangelina had been given a gift from the people of California. Something small, and wonderful: a California cure.

I had the privilege of meeting Evangelina and her family at a California stem cell board meeting.

Evangelina was four years old by then, spunky, wearing a pink superhero T-shirt. She stepped up to the microphone, then jumped back, startled by the squeal of feedback. But she stuck to her task.

"Thank you," she said, in a tiny voice.

Her Mom also spoke to the CIRM: "Thank you — for keeping my family together."

When I talked to her later, she added, "If there is an effort to build Prop 71 part two, count me in!"

In the course of writing this article, I reached out to Dr. William T. Shearer, the doctor who had cared for David Vetter, wondering if he had any comments to make, having given so much of his life to SCID research and therapy. He responded:

"Dear Mr. Reed,

"I know Dr. Donald Kohn very well, and we are collaborators in the stem cell program. Dr. Kohn reconstitutes SCID children with a genetic insertion that corrects gene defect. Many children have benefited from his new form of SCID children reconstitution. I strongly support his research and clinical work with these special patients. Thank you for your note.

"Best wishes for you and your son in 2018.

"Sincerely, WTS." — (William T. Shearer, personal communication)

AND — he reached out to Carol Ann Demaret, Mrs. Vetter, David's mother, who was willing to speak with me.

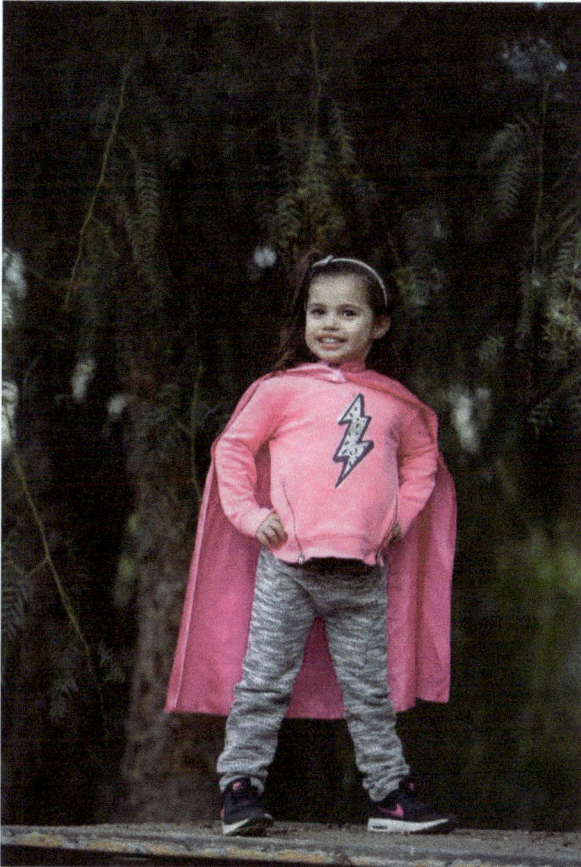

Evangelina Padilla Vaccaro. (CIRM photo.)

I was nervous before our talk. How might it feel for her, to have lost two children to a disease — for which there now existed a cure?

But I "recognized" her right away. She was an advocate, dedicated to the fight against a disease: Severe Combined Immunodeficiency: SCID. She has not quit. She talked about not letting the community down, forging ahead with research — continued help for future generations.

She explained some things I did not understand.

"The bubble was like a second womb," she said, "we thought at first that David's immune system was just late in developing, and if we just gave him a little time, it would activate."

Sustained by strong faith and personal values, she has never given up.

"In the back of my mind, David's loss is always there," she said, "But when people ask me about the pain, I always tell them:

"God sent David to me; research gave us 12 years to spend with him."

After David died, her biggest worry was that his sister Katherine might be a carrier, and might herself give birth to a child with SCID.

When Katherine was 23, married a year, and needing to know, Dr. Shearer called her into his office and told her: she was <u>not</u> a carrier.

Tomorrow, David Vetter's mother will make a beautiful meal — for her two strong and healthy grandsons.

62 Interview with the Founder: Bob Klein

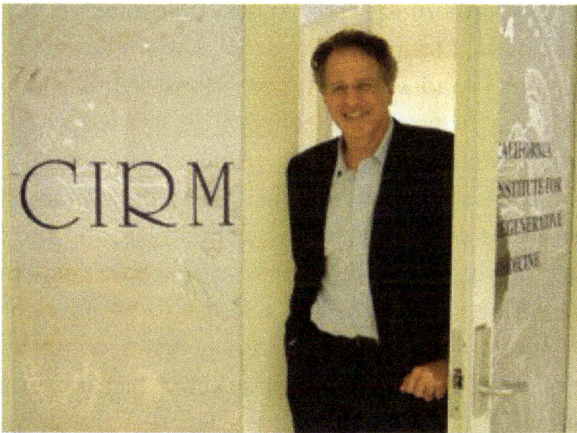

How often do you get a chance to talk to someone who changed the world?

Bob, many people have worked on CIRM, but you are the single irreplaceable leader. The world owes you a debt of gratitude. Thank you.

Let's start with the biggest question: California provided $3 billion for stem cell research and regenerative medicine. How do you answer those who say, "You have already had considerable funding, why should there be any more?"

The fundamental mission was to prove that stem cell therapies could provide an early intervention to chronic disease and injury: to cure or mitigate terrible and tragic diseases that destroy the lives of so many Californians and individuals around the world. The initiative has proven that with Phase 1 and Phase 2 trials. For quadriplegics, it was a tremendous advantage to recover function above the waist. Additionally, for individuals with genetic forms of blindness like retinitis pigmentosa (RP), we have been able to restore functional eyesight in many though not all patients; some have had the condition for so long that it has not been reversible for them at this point. For deadly forms of blood cancers like myeloid fibrosis or polycythemia vera, we know we can restore functional full health.

While we have proven the efficacy of easing disease, the political support is not there in Washington DC to fund the broad spectrum of therapies. We have core examples of how these therapies can restore life and health; we have 2,800 peer reviewed published medical discoveries. This creates a moral obligation since they span the spectrum of chronic disease and injury. We must carry those therapies to the point that biotech companies and foundations can pick them up, whether they are orphan diseases or broad spectrum conditions. While the first initiative (Proposition 71) was to prove the concept, as a discovery initiative, this renewal initiative would be a therapy delivery initiative. It must also continue the basic research because each year the phenomenal scientists in California with the collaboration of scientists around the country and around the world are making extraordinary discoveries like combining genomic engineering with stem cell science to avoid fetal death of a child with genetic defects. (This was actually just accomplished at UCSF in the last 6 months.) We have a remarkable historic opportunity to reduce human suffering. The discovery science is there. The proofs of concept across numbers of diseases and chronic injury are on record. It is our privilege as Californians, Americans, and citizens of the world to support the scientists and doctors who can change the future of families, children, and grandparents; we have a mandate to move medicine forward: in a great leap of advancement.

What are some of CIRM's more significant accomplishments?

As a patient advocate, I would say the most important accomplishment is delivering proof that we can save individuals from tragic deaths by chronic disease, or to essentially lose life through a paralysis that leaves them in a vegetative state. CIRM-funded research has led to the Asterias trials, which now has more than 20 patients with substantial functional return to the upper body. There are currently trials going forward to try and deal with bridging the gap in the spine to the lower body so as to restore function there as well. Whether you use the example of paralysis, blindness, or solid tumor cancer, accomplishment #1 is proving that this is an area of medical innovation that can change human history and reduce the suffering for patients and families. A second major accomplishment is that it has created a body of knowledge in stem cell research and therapy development that is now being coupled with gene therapy to address genetic diseases that can be deadly or radically debilitating. An entire new field piggybacks on stem cell research knowledge to advance the gene therapy field. The third area it has built is the human and physical infrastructure to carry this medical advance forward to access therapies for patients and families across the spectrum of our society. It has funded the training of front line physicians involved with these therapies in alpha clinics set up across the state: creating Centers of Excellence to advance human trials and therapy development from San Diego to Orange County to LA, to the Bay Area to Sacramento. Concurrently, it has built the human infrastructure through funding post docs and grad students who become the next generation, to really bring these therapies to the point where patient access is affordable and broad-based. In summary, the initiative generated the proof of concepts for a broad new wave of historic therapies that would reduce human suffering.

Would you talk about the Alpha site concept and value?

The alpha site concept takes our great university hospitals and creates a network across the state of California for new therapies to be developed and tested and proven through human trials, followed by clinics where the public can access these therapies. These Alpha Clinics lead the way globally by providing a proof for Californians of how to scale these therapies and drive them rapidly with full ethical controls through the universities where they are based. These Alpha Clinics create an ethical umbrella and a confidence that when you go to an Alpha clinic you will be treated with highest ethical standards with the best medicine in the world.

California has official restrictions against state-owned properties. Even so, might it be desirable for an exception to be carved out for CIRM, perhaps as a foundation to be funded by royalties?

The public funding corporation for the initiative, CIRM, has the ability to make loans as an alternative to grants. Those loans when the therapy developed from the grant is successful, the funds coming back to the state agency can fund additional research. They can fund additional therapies and they can fund patient access. CIRM with loan funds has the ability to act like a sustaining foundation, if they put up more of these funds through loans, which I hope they will do.

Do you envision more cooperative involvement with other states and nations?

While I was Chairman we created collaborative programs with approximately 10 countries, some of those programs were very vibrant like the program with Canada or Australia or England. The key concept is that chronic disease or injury does not know any national boundaries, but are plagues on humankind across the world. We have a moral obligation to collaborate with the best scientists of every nation to accelerate therapies and leverage the benefit of the voters' funds from California to advance towards serving patients as quickly and ethically responsibly as possible. While I was Chairman there was $162 million of matching funds for research put up by these countries. In Canada, for example, they put up $40 million for 2 different grants and those grants led to 5 human trials in different types of cancer. It is remarkable that those 2 Canadian match grants could lead in a 7 year period to put medicine on a path to cure these deadly cancers. One of those I know that has come out of UCSD in collaboration of Canada was actually named after CIRM. CIRMtuzumab was going in for registration with the FDA to become publicly available sometime this year or the beginning of next year to the public: for polycythemia vera or myelofibrosis. The prior therapy for polycythemia vera provided symptom relief but could relapse. A close personal friend died because she could not get doctors to understand a prior therapeutic for this cancer was only providing symptomatic relief. The CIRM-funded therapy, however, has a sustained healthy recovery for the individuals.

To me, CIRM is the greatest advance in the history of medicine. I love it as it is, and would prefer to just put more funding in it so it can continue without substantive changes. But for you, as its inventor, do you see the need for change?

I had the privilege of writing the initiative with the aid of a committee of dedicated individuals. With the help of the California voters, we were able to empower the scientists and physicians to drive this revolution in medicine and research. They are the real inventors. As a patient advocate I am a dedicated supporter of their mission. Some of the changes that I am looking for in a new initiative for the voters to consider (along with patient advocacy organizations and Nobel prize winners who previously endorsed) will focus on access and affordability. This initiative is able to build on the foundation and platform of the prior initiative with all infrastructure and discoveries in place.

The opportunity with so many translational grants and human trials advancing is to fund a small group of staff, probably 10–20 throughout the next decade. They will work specifically on access: to see that insurance companies cover therapies for patients, to see that federal programs like Medicare cover the cost of these new therapies; and that the reimbursement for the therapies is not barred by ideologues in Washington, DC. Access means being able to broaden Alpha Clinics further across the state, so individuals that may be remote from one of the Alpha Clinics can find travel funds to and lodging nearby the academic Centers of Excellence.

We need to make certain we have real affordability for those of limited resources. Affordability is more than having an insurance company cover it; affordability means making sure that the pricing and the insurance coverage don't have deductibles insurmountably high. The agency's program must be available to the public, with discounts for public hospitals and clinics. We must have an expert staff with capacity to bring in outside experts, critical in making sure Californians have effective access.

Concern: if CIRM's renewal brings major changes, does that open us to a new threat of lawsuits from those who oppose us for philosophical/ religious reasons?

Fortunately, in retrospect, we were blessed in the lawsuits. The ideological opposition tried to stop the voter's mandate of 7.2 million voters from being implemented. A taxpayer group and religious ideologues who principally drove the lawsuits challenged the initiative on dozens of points. But the trial court, the state court of appeal and then the state supreme court concluded that every single one of the opposition arguments failed. There was constitutional and statutory validation for every part of the California program (approved by a mandate of 59% of the voters), so that we are in the fortunate position that the litigants really could not hold up the programs again. Initially they held up programs from November 2004 through June 2007 except for some funds which Governor Arnold Schwarzenegger advanced to the agency, and some funds I raised by private placement of bonds during litigation. With the decision of the state Supreme Court in May 2007 affirming that the initiative won on every count, there are no real substantive grounds left for suit. There could be frivolous lawsuits but those won't stop CIRM.

You have envisioned California's renewal of CIRM as a "final tactical mile" to go from "initial human trials to broad availability" — could you speak more on that?

I am referring to access. At first the access will be on a practical basis restricted to Centers of Excellence. This will be to make certain that the medical care which individuals and patients are getting is the best: the most tested, highly ethical care that can be provided. It is my expectation and strong belief that once the care is proven through all Centers of Excellence and the physicians who associate with the Centers of Excellence, there will be a broad spread of these therapies to all of the major hospitals in the state. This is a process that will not be accomplished overnight but certainly over the next 10 years. If the voters choose to approve this renewal of funding ($5.5 billion), within the next 10 years after approval I would expect this revolution of medical care to be broadly shared across the state.

We now have Induced Pluripotent stem cells, reprogramming, and other forms of making cells for research and cures. Do we still need embryonic stem cell research?

Embryonic (also called pluripotent) stem cell research and therapies have been a vital contributor to the progress. Asterias Therapeutics' clinical trials for quadriplegics is derived from embryonic stem cells that create a special class of neural cells call oligodendrocytes. Without embryonic stem cell lines, that therapy would not exist. Age-related Macular Degeneration (AMD), a common form of blindness, is being fought by 2 separate teams, based on embryonically derived retinal cells. AMD will lead to partial or full blindness for 1 out of 4 people between 65–85. Imagine working your entire life and then retire, only to find you have gone blind; you cannot drive a car, cannot see friends. AMD therapy is derived from embryonic stem cells. Those are just 2 examples of critical breakthroughs for tragic areas of chronic disease or injury that are dependent on embryonic stem cells. Broadly, embryonic stem cells show us the evolution of differentiation and development of cells. ISSCR

(International Society of Stem Cell Research), representing the best scientists across world, said in December of 2018 that hESC is the gold standard for therapy development and research. Embryonic stem cells are derived from excess cells thrown away from In Vitro Fertilization (IVF) procedure. When a family does not want any more children, they may choose to have excess IVF cells frozen, thrown away — or contributed to future therapy. Orrin Hatch, a pro-life republican senator who just recently retired, said it was definitely a prolife form of medicine to use these cells that would otherwise be medical waste — to help save lives or restore quality of living to patients. I think he hit it on the head.

Are you concerned about future or present threats to regenerative medicine? Is "personhood", favored by vice President Mike Pence, a continuing threat?

We know the state government of Indiana has prohibited embryonic and fetal tissue research, based on the theory that any fertilized egg is a person. Fundamentally, this is a religious decision for the individual to decide. Certainly, the voters in Mississippi and Colorado rejected similar personhood initiatives by huge margins. They understood that if laws were enacted that prohibited doctors from developing therapies from cells or if you were required to throw cells away that were fertilized or designated as medical waste you would be in criminal violation of such "personhood" laws. Voters understood that it would forever interfere with IVF, a procedure that has given birth to more than 6 million people. I have faith that when the public gets the right information, they understand that personhood laws would really interfere with a family's ability to have children through IVF. It is vital that each person respects democracy and policy governing medical research. These are decisions to be made between families and their doctors, not imposed by ideological constraints.

You have mentioned the importance of "early intervention strategy" as needed to prevent infant deaths — could you say more about that?

Early intervention strategies apply to many medical situations. When an infant is diagnosed with a deadly genetic condition like thalassemia, or an individual suffers a stroke, they may need immediate stem cell therapy to regenerate damaged tissue, resolve a scar, or rebuild the damaged brain. It is critical to intervene quickly in blood cancers like myeloid fibrosis before there is compromising damage; before the person is irreparably harmed, and before society has been burdened with the endless support of an invalid, and a family bankrupted by years of expensive therapies. This means paradigm change: to reduce or eliminate the cost of long-term therapies.

Inspired by CIRM, six other states now have stem cell programs, but all have funding difficulties. Do you have a message for them?

One of every 8 Americans lives in California. After the first initiative (Prop 71), a number of states launched programs to advance stem cell research and therapies on a smaller scale than California. Whether it was Maryland, Washington, Connecticut, New York, Texas or Wisconsin, each state had critical contributions to make. At the end of the 2004 initiative, in the very early days of the internet and social media involvement, there were 3 million affinity group emails generated in California. These came from voter families and societies to families and friends talking about the importance of this research. Also about 3 million emails went outside the state. All that support and voter approval in California (59% said yes!) influenced the politics in other states. I believe that the second initiative in California, if strongly supported by voters, (with 60% or 70% approval), and resources of $5.5 billion, will encourage other states to participate in this renaissance of research.

Chronic disease is everywhere. I cannot imagine a higher priority than reducing sufferings that afflicts individuals and families of 40–50% of all Americans, especially toward the end of life. Here is a great opportunity to enhance quality of life for very extended periods for senior citizens in California and every state. When that opportunity becomes clear in California, voters in every state will demand participation, both locally and through federal support.

Bob, long ago, I think I remember you mentioning a historical precedent (the Medici?) for your invention of CIRM?

The Medici family in Florence made a great contribution to science by taking a stand against religious ideology which was then supposed to dictate every scientific answer. They created and funded a scientific society based upon empirical evidence from experiment: creating effectively an academy of science. That academy later gave birth to a national academy in England, the Netherlands, in Spain, and eventually in the U.S. The Medicis protected science and the fundamental empirical study of the human body from religious doctrine; which claimed at that time to have the answers to all scientific knowledge, denying the need for experimentation. We are again at a historical moment when medical science can only survive if every citizen, civic group, Nobel prize winner, philanthropist, indeed every person committed to health will stand together against ideological pressures to shut down research. We need support through an active and engaged citizenry. Like a modern Medici family, but living in a democracy, we have a responsibility to protect that tradition of freeing science and medicine from ideological censoring and bans.

As you know, President Trump just banned federal funding of fetal tissue research. It is critical that the public understands that all the way back to the late 1950s, fetal tissue has been critical to families in the U.S., starting with therapies for Polio. The only way that the vaccine for Polio could be effective then, by multiple dosages, was to nurture that vaccine on fetal tissue at the Windstar Institute in Pennsylvania. Alternative methods of propagating that vaccine failed. There was a huge epidemic of Polio ravaging the country. Plans were drawn up by government to have subsidized hotels to house vast numbers of people with Polio.

Ironically, some of the individuals that are banning this fetal tissue research today, actually benefited from it themselves. It protected them from Polio. But yet they are today denying this approach to research in building new vaccines. Why is that just? Why is that fair? Why would we shut off an area of science that has been so important to us? Polio was just a start. Fetal tissue was used to develop vaccines for tetanus, adrenal virus, hepatitis A and B, MMR vaccine, rabies, chickenpox and more. Why is it with all of the medical benefits that fetal tissue research has given us, in protecting hundreds of millions of Americans and individuals throughout the world, we would now ban the tissue that was the foundation of that research — and which is still critical to ongoing research? I don't believe there has been enough public education and understanding of the ethical protections under which this research is conducted.

This ban that just came out reportedly is killing a grant right now, today, at UCSF for research towards a human trial of AIDS. Why would we stop research to cure individuals from AIDS? Our constitution provides for a separation of church and state. Morally I do not believe it is right for me or anyone else to ban medical research that is critical to another family's children, spouse, parents, brothers or sisters; everyone should have the right of access to the best medical research and medical therapies available.

Dr. Paul Berg led a revolution of incredible importance in the medical history of this country and the world. He was awarded the Nobel Prize with Dr. David Baltimore, both Californians, for recombinant DNA research. That research led to 100 critical heart and cancer drugs just in the first 3 decades since its discovery. That has saved the lives of literally millions of Americans. The research was revolutionary and during initial stages it was opposed by religious ideologues who did not agree with it. Protestors shut down the Harvard labs in Boston claiming God had not willed mankind to have such knowledge. In Washington, DC, congress voted in more funds in 1978, and great discoveries were announced. The City of Hope and UCSF jointly announced creation of artificial human insulin that kept my son alive for 15 years.

I deeply embrace this lesson in history. When Dr. Berg was taking me to meet with a funder who he believed would donate to our initial campaign, I was riding in the car with him. I remember saying, "Dr. Berg, your research created a revolution in medicine. Do you think the stem cell therapies will ever be even close to as important?" Dr. Berg looked at me and said: "Bob, you just don't understand. This therapy approach with stem cells is going to change the future of medicine in California, in the country, in the entire world. In 20–30 years you won't even recognize how medicine was ever practiced. This is far more important in its global impact on medicine and human suffering." Dr. Berg's statement left me speechless, and for the rest of the half hour ride in the car I only thought about the moral obligation to pass the first initiative, which fortunately the voters of California embraced. The voters of California started a revolution of discovery, 7.2 million votes for Prop 71 which (by the way) was at the bottom of the ballot after everything else. It got as many votes as the U.S. Senator at the top of the ballot who got the greatest number of votes of anyone in that election. The public understood this was a medical revolution that could affect every family. Their vision was right and the new revolution of therapies will (as Dr. Berg foresaw) change medicine forever — especially now that it has empowered genetic research and genetic stem cell research and therapies can move forward to address chronic disease and chronic injury as an unbelievably potent pair of medical answers to ageless suffering.

Finally, a great question from Melissa King: "Bob, what resonates most with you, 15 years later, as you talk with scientists and patients that have interfaced with CIRM, with what you wanted Prop 71 to accomplish?"

I would wager that 19 out of 20 scientists in the biological sciences in California who interacted with CIRM programs believe this is a critical bridge to the future of medicine and science. All of this phenomenal knowledge, new therapies and lives saved, was made possible by a visionary group of voters. They turned out to the polls, and they voted their conscience. They understood the best future they can dream of achieving was not the history of a broad public suffering from chronic disease and injury, but a new future where health is sustained into very old age with good function and vitality, where one can enjoy life with their families.

Fundamentally, I believe in the California voter. Given full information, they will vote for revolutionary medical therapies. My mother died in her 90s from Alzheimer's. In the last decade of her life she lost all memory of her children, all the charity groups she contributed to, her friends, her family in Missouri, everything. It was as if the world shrunk to a very small place that only occupied the ground she stood on and she didn't know anything outside of that space. That is not the future I would dream of for her. Instead, our goal should be a healthy life right up to the very end. Certainly, stem cell and genetic therapies will not cure all diseases immediately. But stem cell and genetic therapies will so broadly improve health and human life that they will change the future of how we live and how we contribute to society. They will free the state and country, family and individual, from massive medical debts. Our dreams will not be perfectly realized, but they are broadly achievable in the medical field. I wish they had been there early enough for my youngest son, but they will be there for hundreds of millions of other children and families. Getting to these ultimate goals will take a long time but making substantial progress can be achieved within the next 15 years. The quality of human life can be greatly improved.

Importantly, pushing back the cost of chronic illness will free up money we need for dealing with climate change, for improving the quality of education, raising teachers' salaries, addressing hunger and more. These therapies can bring health to the economics of our state and nation while restoring dignity to the individuals and families of our state, our nation and our world.

63 The Big Bang Theory, CIRM, and a Dolphin Named Spock

After twelve years of audience delight, science-supporting 'BIG BANG THEORY' will make no new episodes... (Entertainment Weekly photo.)

It is hard to believe this is the end of my favorite TV show, THE BIG BANG THEORY. I own the DVD set of "BANG," all available episodes (11 seasons currently, and the 12th is on order). I know the show so well, I often shout advice to the folks onscreen.

"No, Leonard, don't say anything!" I yell when he (Johnny Galecki, self-proclaimed King of the Nerds) and Penny (the impossibly gorgeous Kaley Cuoco) are in bed together for the first time — and Leonard gets in trouble when he ignores my advice...

I wish the producers would bring together some of the best stories (Leonard and Penny's romance, Howard's space adventure) so they could be enjoyed separately.

But the most inspirational moment? During the last show, Amy Fara-Fowler (Mayim Bialik) gives a roaring shout-out to girls and women who want to become scientists. Owner of a real-life PhD. in neuroscience, Bialik knew whereof she spoke, and she nailed it, wringing actual sobs from at least one viewer.

Every science classroom should have a copy of that speech framed on the wall!

Writer/Producers Chuck Lorre, Bill Prady and Steven Molaro deserve a thousand thanks for this hilarious and touching show.

Gloria and I feel sad that the Golden State Warriors are leaving Oakland, but let me tell you a story to take the sting out of the relocation; it concerns a dolphin, and a Golden State Warrior...

I was working as a diver at Marine World, and one of "my" divers (meaning I hired him, you will see why I claim him in a moment) named John Racanelli was repairing a filtration pipe in the dolphin tank.

A dolphin named Mr. Spock got too close to one of the stainless steel bolts — and *swallowed* it.

We set up a net in the dolphin tank (not to catch Spock, but to narrow his swimming area) and swam down into the transparent water.

Sometimes the dolphins fought us when we caught them for medical attention, but I went slowly, gradually, letting him get used to the idea. I was holding my breath; the air hoses we normally used ("hookah lines") could tangle us up in a catch.

There he was, Mr. Spock, right at the bottom, looking up.

My arms slipped slowly around his smooth-muscled 500 pound dolphin body, and he let me take him. I swam him to the surface, where the stretcher was waiting.

We carried the dolphin down to the vet clinic, and put him on the examination table for the vet. Gently, carefully his jaws were separated. Vet Ron Swallow slid his greased gloved hand down the dolphin's throat, into his first stomach, reaching...

I held Spock's tail, with difficulty; he was not enjoying this.

"I lost the bolt," sighed the vet," It slipped out of my fingertips, into the second stomach." (Dolphins have three connected stomachs.)

Everyone got worried. Would the vet have to do surgery on a dolphin? Since dolphins do not breathe automatically, but must "ask" for every breath, surgery is complicated and dangerous for them.

And then the publicity lady, Mary Jo O'Herron Ball, said the magic words: "I know a Golden State Warrior..."

An hour passed — and then an amazingly tall man came entered the vet clinic.

His forearms had to be shaved. Then he gloved up, slathered on lubricant. We held Spock still once more as the professional basketball player reached slowly in, way in, till his shoulder almost touched the dolphin's jaws...slowly, slowly...

"You have to pull out, you've been in there too long!", snapped another vet, Jay Sweeney, listening in on the phone.

"Got it," said the giant, and removed the shining deadly bolt.

And that was how Golden State Warrior CLIFFORD RAY became the hero of Marine World, for saving a dolphin's life.

The diver, John Racanelli, went on to become President and CEO of America's magnificent National Aquarium, in Baltimore, Maryland. He has since designed what I consider the perfect way to view dolphins, in near total freedom for them...

It is called the "Dolphin Sanctuary" and will open in 2021 in Puerto Rico.[1]

As for Marine World, it was almost destroyed, for financial reasons. A $7.5 billion multi-national corporation bought the land underneath us.

But we, the employees of Marine World, to us a little patch of Eden, did not accept that verdict. I remember at the very meeting the corporation held to announce its intention of turning Marine World into a mall, I spoke on behalf of the employees, saying:

"We will fight you..."

---

[1] https://www.youtube.com/watch?v=y9Hd8evW9BM

And we did, rallying the community. The corporation changed its mind. I was actually honored when the corporation insisted that my name personally be on the agreement, that we would accept their changed offer (16 months before we closed instead of 4), allowing us time to relocate instead of shutting down.

It was my first political battle. Everyone told us there was no chance, but we won anyway.

To this day, if you drive to Vallejo, California, you will find an entertainment center called "SIX FLAGS" — and half of it is sharks, dolphins and tropical fish...

And CIRM?

As you know, the California Institute for Regenerative Medicine is a shining place for me and millions, ever since the first campaign for it began in April 2003. There is a lot of talk about an end of CIRM, but some of us are not prepared to accept that.

I love to attend the meetings, and was recently asked to speak at one.

Here is what happened, as recorded by champion transcriber Beth Drain.

Board Chairperson Jonathan Thomas said:[2]

"...a number of us have been interviewed recently by the great chronicler and prolific author Don Reed. And I thought it might be nice for Don to spend two or three minutes to talk to the board about his project that he's working on, and the upcoming book that will result. So, Mr. Reed, will you proceed to the podium?

Senator Art Torres: "When will <u>Jensen's</u> book be done?" (David Jensen, author of the massive weblog, "California Stem Cell Report").

Reed: "I've been trying to get him to do one! (And he is!).

"There's a lot of talk about will there be a Part 2 (of Prop 71) initiative, and I'm not the person who can answer that. I remember what Bob Klein said about it, though. He said there will be a poll, and on the result of that poll he will make the decision. For my mind, there's never been the slightest doubt. The work you all do here is irreplaceable. It's magnificent and it changes the lives of so many.

"The next book will be titled: "REVOLUTIONARY THERAPIES: How the California Stem Cell Program Saved Lives, Eased Suffering, and

---

[2] https://www.cirm.ca.gov/sites/default/files/files/agenda/transcripts/Transcript%20 ICOC-ARS-5-23-19.pdf

Changed the Face of Medicine Forever". That's what has been done. This is something incredible. It will be the greatest thing we will ever do, in all our lives.

"The book is due out by Christmas, and that means my part has to be done by July 29. So I'm scribbling frantically as we speak.

"It was a delight to interview President Millan, Chairman Thomas, Maria Bonneville, Kevin McCormack, Bob Klein — but you know, every connection that I have had with this organization has been positive and to be remembered...

"So that's pretty much it. I just scribble, scribble, scribble.

"But I would like to tell you one little story...

"The great singer Al Jolson would always perform free for charity, but he had one condition... he would only do it if he was the last person that sang, in the star position. And so one day the fund-raisers came up to him and said: 'We want you for a really important event — at Carnegie Hall! But there is one little problem — you don't want to be the last person to sing this time.'

"Al Jolson said, I have to be. That is my condition.

"And they said, 'Do you know who is going to be singing, the last person?'

"He said, 'No, who?'

"And they said: 'Enrico Caruso.' (At the time, the greatest opera singer in the world.)

"And Al Jolson said: 'My condition stands.'

"So came the night, and Caruso came out first, and he sang — everything: Vesta LaGiubba, Ave Maria, everything. He sang and sang and sang. The people would not let him go. They kept him there till he was exhausted. Their hands were sore from clapping so hard.

"And then Al Jolson bounded out, beaming that wonderful smile.

"And he said, 'Hang on to your hats, folks — you ain't seen nothing yet.'

"Well, folks, that's how I feel about CIRM: the California Institute for Regenerative Medicine. You 'ain't seen nothing yet.'

"The best is yet to be."

64 For My Son

Roman Reed
President, Roman Reed Foundation
patient advocate

And — Roman Reed just announced his bid for a seat in the California Assembly! (CIRM photo.)

As you know if you read any of my previous books, my son Roman Reed became paralyzed in a college football accident, September 10th, 1994.

For two and a half decades, he has suffered the agonies of the damned.

Roman is brave, and does not share his troubles with outsiders. But I am his father, and I see what he goes through: the skin breakdowns, infections, circulatory problems, and all the endless frustrations of paralysis, in an able-bodied world. But his attitude remains like the lions of Tsavo, which were so fierce they shut down the building of a railroad (and inspired the world's first 3-D movie, "BWANA DEVIL").

Two days after the accident, his sister Desiree found out about a clinical trial (Sygen, an injectable compound made from dried cow brains) and I tried to get him included. Unfortunately, it was raining the night before the trials began, and the head nurse advised us against bringing him in during the storm — the morning would be fine, she said. But it was not fine. We missed the deadline by one hour.

Obtaining the Sygen from Switzerland, we injected Roman ourselves (his girlfriend Terri did most of the shots) and got him into a "rehab clinic" of which I will say nothing. We made a gym out of our apartment, and he exercised every night. How does a paralyzed person exercise? The care giver moves the arm or leg, while the paralyzed person helps as best he can. It is exhausting for all concerned.

But whether it was the Sygen or the exercise, Roman recovered the use of his triceps, and learned to drive an adapted van.

We passed a research-funding law named after him, the Roman Reed Spinal Cord Injury Research Act of 1999. On March 1, 2002, a rat which had been paralyzed walked again — thanks to Dr. Hans Keirstead and an embryonic stem cell therapy. You might have seen a 60 MINUTES TV presentation about that.

One curve followed another, up and down, good and bad: Geron Corporation took over the product, and wrote some 20,000 pages of documentation on it, corresponding with the FDA.

Then new Geron management took over — and killed the project.

Good news followed bad. Ed Wirth and Asterias Biotherapeutics bought the therapy, and with financial grants from CIRM, it began to move again.

Twenty-five injured people received doses of embryonic stem cells, in varying amounts. All achieved the safety goals "doing no harm". In addition, most recovered upper body improvements, like hand and arm control.

"One year after being treated, all the patients are doing well, none have experienced any serious side effects, and most have experienced impressive gains in movement, mobility and strength."[1]

One of the pioneering recipients of stem cells was Jake Javier, who received ten million cells, half the full dose. That sounds like a lot, but it isn't. This amount was not expected to do anything, just to test if it would be safe, which it was.

---

[1] https://blog.cirm.ca.gov/2019/05/03/one-year-later-spinal-cord-therapy-still-looks-promising/

I spoke with Jake, a cheerful person with movie star good looks. He had received his injury in a diving accident. Jake expressed appreciation to CIRM, all the scientists like Hans Keirstead who developed the therapy, and Asterias Biotherapeutics, who brought it all together.

He achieved some triceps function return, and a little bit of control of the right index finger. Jake continues college, working toward a degree in bioengineering.

Naturally, I wished Roman could have had the stem cells too.

But the therapy was only for new injuries ("acutes") — and Roman was "chronic", like anyone who has been paralyzed more than a few days.

The problem (remember?) is the scar. Made by the injury, scar tissue blocks messages between brain and body. The scar must be removed — but how to do that, without increasing the injury? We did not want to make things worse.

The years passed by…

Roman built connections in the SCI community, meeting people like no one else. If he heard about a newly injured person, he would drive half the length of California to chat with him or her, let them know about the fight being waged on their behalf.

And every day Roman's paralysis remains, with no days off, ever.

But I see two new reasons for encouragement.

One of this book's chapters (on Rett Syndrome) was built around a young Chinese scientist, Dr. Yi Eve Sun. She was also working on spinal cord injury with a scientist named Dr. Xiaoguang Li, at the Beijing Advanced Innovation Center for Biomedical Engineering, Beihang University, Beijing, China.

I was delighted to learn that their project began with the *removal of the scar*. They used an MRI device to "guide the excision of scar tissue."[2]

After that, a gel called NT3-chitosan (a nerve fertilizer) was placed in the wound.

"NT3-chitosan regenerat(ed) the chronic spinal cord injury…Taken together, our study demonstrated (the) feasibility of using NT3-chitosan transplant to enable SCI repair and (MRI) to guide scar (removal) and

---

[2] Rao J-S, Zhao C, Zhang A, *et al.* (2018) NT3-chitosan enables de novo regeneration and functional recovery in monkeys after spinal cord injury. *Proc Natl Acad Sci U S A.* 12; 115(24): E5595–E5604. https://europepmc.org/articles/pmc6004491

monitor regeneration, potentially in patients entering...the chronic phase."[3]

That paper, "MRI-guided scar removal and regeneration elicited by NT3-chitosan to treat chronic SCI" — was one reason I had struggled years to learn Mandarin. Even though I failed, I can still at least show courtesy to scientists from the Middle Kingdom. I can say, "Wo shuoda bu hao" (I speak very badly) and "gan xi bao shir she de hen hao" (stem cells are very good!) to Chinese scientists, and fortunately they are patient with my poor communication.

If California achieves the hoped-for renewal of funding ($5.5 billion proposed), I hope Drs. Li and Sun and their fellow scientists will set up a branch in California, so they can apply for grants, and try to cure chronic paralysis.

That's one piece of good news, here's another:

That therapy was later bought and sold by a series of companies, ending up in Lineage Cell Therapeutics. Just days ago, the company attained a patent on the procedure, using precursor oligodendrocytes to re-insulate damaged spinal nerves. They will be using OPC-1, the original stem cell line developed by Hans Keirstead on a Roman Reed grant.

And maybe one more ...

Recommendation for TRAN1-11579, Public Comment by Don Reed:

"Honorable Members of the ICOC: Thank you for allowing me to make public comment on Application # TRAN1-11579, the Human Embryonic Stem Cell-Derived Neural Stem Cells for Severe Spinal Cord Injury.

"More than twenty years ago, Dr. Mark Tuszynski was one of the first recipients of a Roman Reed grant for spinal cord injury research. Ever since, he has been working, quietly and steadily, in tandem with the field of regenerative medicine.

"His project today, TRAN 11579, is the culmination of that lifetime of hard work science. And it is spectacular. Usually, when I see a photograph of a spinal cord injury, it is hard for me to understand what is going on. It is a complicated X-ray of the spine, and the scientist points to a v-shaped little notch and says, that is the injury — and see that little fuzzy growth on the edges of the wound? That's it, that is the regenerated nerve. And I nod my head and smile, but it takes a lot of faith to see anything.

---

[3] http://www.xinhuanet.com/english/2018-05/31/c_137220891.htm

"Dr. Tuszynski's work is different. It is impossible NOT to see new growth — green-marked nerves leaping across the gap in the injured spinal cord, bridging to the other side. The new nerve cells were biomarked with green so you could follow the growth — it was like the whole spine was slathered with lime green paint.

"And what did that mean in practical terms, was there any recovered motion?

"'Injured rats, with completely severed spinal cords, regained significant motion, including the ability to move every joint of their legs,' said Dr. Tuszynski.

"That was rats. He has since gone on to achieve similar results with a non-human primate, the rhesus monkey.

"Nature Magazine has honored him by publishing no less than six articles on his recent work, articles like: 'Restorative effects of human neural stem cell grafts to the primate spinal cord,' and 'Biomimetic 3D printed spinal cord scaffolds for spinal cord injury,' and, most recently, 'Chondroitinase improves anatomical and functional outcomes after primate spinal cord injury,' which is in press.[4–6]

"He has done every step of the early work required, achieved strong preliminary results. It is vital that his work go on, and I urge his continued support.

"And maybe my son will benefit, he and so many paralyzed people, like Karen Miner, Susan Rotchy, Fran Lopes, and more — after all the years of everybody's hard work."

And the vote was to approve ...

---

[4] Rosenzweig ES, Brock JH, Lu P, *et al.* (2018) Restorative effects of human neural stem cell grafts on the primate spinal cord. *Nature Medicine* 24:484–490.

[5] Koffler J, Zhu W, Qu X, *et al.* (2019) Biomimetic 3D-printed scaffolds for spinal cord injury repair. *Nature Medicine* 25:263–269.

[6] Rosenzweig ES, Salegio EA, Liang JJ, *et al.* (2019) Chondroitinase improves anatomical and functional outcomes after primate spinal cord injury. *Nature Neuroscience* 22:1269–1275.

65 The ISSCR Adventure

The International Society of Stem Cell Researchers (ISSCR) is a chance for champions to talk with each other and the world — intellectual warriors. New President Deepak Srivastava promises it will be fun. (CIRM photo.)

Gloria had another heart attack, just as our plane was rumbling down the tarmac.

"The pain is wrapping around my chest and back," she said.

Fortunately, Gloria always carries a bottle of nitroglycerine with her nowadays, and popped two of the tiny white cubes under her tongue.

"It burns," she said.

"Your heart?"

"No, my tongue," she said, "from the nitroglycerine."

"Should we stop the plane, and go back — ?"

"To go to the hospital and get observed? Have somebody tell me I should exercise more and eat less? No, it is calming down," said Gloria, and so we continued on our way to the International Society of Stem Cell Research (ISSCR) conference.

The ISSCR is a huge organization, with 4,000 plus members representing 60 countries across the globe.

The venue for this year's conference, the Los Angeles Convention Center, was like something out of ancient Greece, shining white marble, fountains of leaping water and the (somewhat anachronistic) huge escalators.

One of the best parts of a conference is shaking hands with scientists you only knew as names before.

Like bearded Pete Coffee, now at the University of California at Santa Barbara, (UCSB) but originally from the UK. He was attacking blindness (Age-related Macular Degeneration, AMD), and of course I had questions.

What had he done with his CIRM leadership grant?

"We tried to duplicate the London Project to Cure Blindness," he said, "Working with Mark Humayun, Dennis Clegg and David Huxton."

What was his approach to fight blindness?

"An eye patch," he said, "like repairing a bicycle tire. Put a layer of stem cells on a sticky plaster, add that to the back of the eyeball."

Did it work?

"Our first patient could only read one word a minute. Today, four years later, he can read 80 words a minute. Our second patient could not even see the book at first — now he can read 50 words a minute."

What did he think about CIRM, invented by Bob Klein?

"Because of CIRM, we are restoring biology to a level where people no longer just exist, but actually live, because of the treatment.

"In the UK, we would give Bob Klein a knighthood, call him Sir Bob."

As Gloria and I sat in the audience, I had a feeling of a world turned full circle.

In 1996, when Roman had been paralyzed just two years, I attended my first conference on regenerative medicine, in Asilomar, California. I asked so many questions, I was told: "No more, or you will have to leave."

To say the science was over my head? That would be too kind. Today, I can follow if the scientist talks slowly, and doesn't use too many big words. But when a lecturer gets excited, and talks faster and faster, like an encyclopedia on speed, every word longer than the one before... even now my head begins to nod.

At the ISSCR conference, the scientists were very excited indeed … Stanford's Anthony Oro is a brilliant scientist, working on a way to keep skin from breaking apart. The title of his speech was "Chromatin dynamic strategies during surface ectoderm commitment" — which means, pretty much, ways to keep skin from breaking apart! But with him, as with most scientists, you just have to keep skin asking him, "what does that mean?," or "could you slow down a bit, please?"

Gloria had brought pain pills (the shoulder operation) and sleeping pills, and during one of the speeches, unfortunately, she took one instead of the other…

But it worked out. When one of us would start to snore, the other would just nudge him or her, two sleepy senior citizens, trying to stay awake!

Impressions:

The young and the old scientists seemed equally represented; Barbara Treutline won the Susan Lim Award as most promising young scientist — among her studies was the salamander-like axolotl, which can regenerate its limbs. Interestingly, when one of its legs is missing, the non-injured limb appears to send messages, helping the absent leg to regrow.

Among those with a more mature age-range was craggy-faced John Gurdon, who had earned his Nobel Prize (on Somatic Cell Nuclear Transfer, the cloning of cells) the same year Shinya Yamanaka did. With his swept-back white hair, Dr. Gurdon embodied dignity and grace.

One scientist I personally wanted to succeed was Arnold Kriegstein, of UC San Francisco. Not only is he working to reduce the seizures of epilepsy, but also to control neuropathic pain, like that which pains my feet.

A presentation I regretted missing was the "Women in Science Luncheon". Unfortunately, advance registration was required, grumble, grumble. But talk about an important subject! With more than half the world of the feminine persuasion, we dare not deny ourselves the participation of women scientists — sexism is a stupidity we cannot afford.

Jens Puschhof of the Netherlands had an unusual topic — making organoids out of snake poison. I am not sure why he wanted to work with snake venom glands, except perhaps that 2.7 million people are bitten by poison snakes every year?[1]

---

[1] https://www.who.int/news-room/fact-sheets/detail/snakebite-envenoming

Karen Aboody stepped up for stem cells, representing the City of Hope's efforts against cancer.

The Parkinson's community showed real strength, especially New York's Lorenz Studer and his Blue Rock company, reportedly funded to the sum of $225 million...[2]

Stanford's Irv Weissman always seems to have something new up his sleeve, including a possible way to remove the need for chemo and radiation treatments...

I remember a time when only Helen Blau of Stanford seemed to be working on problems associated with aging, but that was certainly not the case anymore, with people like Weigi Zhang of the Beijing Institute of Genomics working on the "arterial aging of the cynomolgus monkey"; Nalapa Reddy of the Cincinnati Children's Hospital, on the "Amelioration of Intestinal Stem Cell Aging"; and Romeo S. Blank's efforts on "Aged macrophages and...muscle regeneration," supported by the University of Rochester, New York, to name a few.

A surprise to me was how the zebrafish can regenerate its fin, a fact which may be useful in the healing of our own ligaments and tendons, according to Jenna Galloway of Harvard Medical School.

Zubin Master of the Mayo Clinic spoke on the importance of educating patient advocates (like myself) on "evaluating...unapproved stem cell treatments..." This is vital, lest people pay many thousands of dollars for what may or may not be stem cells, not only risking being cheated, but possibly being exposed to danger.

Doug Melton of Cambridge came up with an IPS model of Type 1 diabetes, and a way to "boost the yield of insulin." To say that Melton is respected in the field is like saying Mt. Everest is a hill, and this appears to be the culmination of many years' of effort. How important? Around the world, an estimated 422 million people have diabetes.[3]

Doug Sipp of the Riken Center (Japan) took on the issue of pricing stem cell therapies. This problem must be solved; what good is an unaffordable cure?

How should scientists communicate with the media? Several experts shared sage advice: Sean Morrison, Sally Temple and Alan Trounson

---

[2] https://ipscell.com/2017/11/on-the-threshold-of-cell-therapy-for-parkinsons-disease/

[3] https://hsci.harvard.edu/news/related-faculty-member/douglas-melton

gave the scientists' side. Bradley Fikes (San Diego Union Tribune) and Meghana Keshavan (USA TODAY, STAT) clarified the needs of the press.

Canada's John Dick received the ISSCR Award for Innovation, based on his pioneering efforts to fight leukemia.

Billionaires Eli and Edith Broad were lauded for their incredible gifts to the field of regenerative medicine. It is difficult to find any corner of research which has not benefited from their generosity.

Coolest name for a scientist? Frederic de Sauvage of Genentech, California. But Dr. Sauvage is a lot more than a name. He is working to develop models of colorectal cancer progression, one of the deadliest forms of disease.

Marcus Grompke, of Oregon Health and Science University, is working on a way to regenerate the liver. I thought about my cousin Will and his wife Mary, and their cross-country quest to find a usable liver, to help keep her alive, a fight which continues to this day.

One scientist told me she was from Washington State. I asked her did she know Washington State had a stem cell funding program, which might help research projects like hers? Charles Murry helped author the newest stem cell research program in America...

Ryuichi Nishinakamura of Kumamoto University, Japan, is attempting nothing less than "Building the kidney...from...stem cells."

Deepak Srivastava (Gladstone Institute) gave a rousing five-minute speech on "Engineering Tissues and Organs," during which he promised that his tenure as new President of the ISSCR would "always contain an element of fun"...

Clive Svendsen, of "Answer-ALS," is leading a gigantic (1,050 participants!) effort to "Identify ALS subgroups, biomarkers and druggable pathways... Whole genome sequencing was conducted on all participants. In addition, a smartphone-based system was employed to collect deep clinical data including fine motor activity, speech, breathing and linguistics/cognition..."[4]

Larry Goldstein of UCSD sat quietly in the back of the room. Not only is he a champion of science, but someone who can write clearly and with impact — Dr. Larry was given a Lifetime Achievement Award for his contributions to the field.

---

[4] https://www.answerals.org/team-member/clive-svendson-phd/

Silver-haired Bernie Siegel did not get an award, but deserves one. His long-running World Stem Cell Summit does more than anything else to bring the world together around the various facets of regenerative medicine. As this is written, his next event will be Jan 21–24, 2020, in Miami, Florida.

In addition to his speech before the full audience, Bob Klein spoke privately to the attendees from Americans for Cures Foundation. I am sworn to secrecy...

I was glad to shake hands again with Pat Olson, former Chief Scientist for CIRM. Sadly, she is retiring, but (happily!) will continue to serve on a consultant basis. Her knowledge and practical experience are irreplaceable.

And there was Arlene Chu, who keeps saying she is retired, but is somehow always where the action is...She had a big scar on her knee, and told Gloria it was from a knee surgery. How bad did it hurt?, asked Gloria. "The pain was worse than having a child," said Arlene, "Everyone told me it was not so bad, but..."

People from Neurix (a biomed company) spoke on the importance of organoids, which might save the field billions by pointing out bad products early...

Others spoke on the importance of chimeras, animals which could grow human cells or organs inside them, a potential assist for people.

The advocates for Americans for Cures Foundation were taken out to dinner at a place called "Public Schoolhouse" — serious fun.

Adrienne Shapiro, sickle cell advocate par excellence, and her daughter and husband were on our left. Adrienne and I got to talk research funding politics, which of course I enjoy.

Across from Gloria was the Vaccaro family, famous for Evie, the girl who beat the "bubble baby" disease, with the help of science. Earlier she played "Battleship" with an advocate, and I am reasonably certain it was an accident that she spilled her milk on his shoes.

But now she was tired and leaned her head against her father, Christian, and the affection was very plain between them.

"She likes to comb my hair and put ribbons in it," he said, shaking his head, while she grinned and snuggled in deeper. For us, Evie was a shining star, summarizing all our hopes and dreams of cure. But in that moment, she was just a little girl, tired of all the talk, and just wanting to go to sleep.

Do you know Bob Klein's son, Robert? He is very kind and mannerly, but also quietly determined, a person you want on your side for the long battle.

Our own Jason Stewart, director of advocacy for the Americans for Cures Foundation, spoke on the practical elements of getting an initiative on the ballot, as did Paul Mandabach, of Winner and Mandabach, so crucial to the Prop 71 effort.

I spoke to the patient advocates, which was a joy. I saved the speech so you can read it, if you like, in the next chapter.

Unfortunately, Mitra Hushmand's speech was cut off by pressures of time: a serious loss. If you get a chance to hear Americans for Cures' Director of Science, be sure to take it: a personable speaker with something to say.

Field Coordinator Melissa King is a ball of fire, always on the lookout for a way to advance the effort. When Gloria and I could not figure out "Uber" (or they could not find us) she stayed outside on the sidewalk with us, pushing phone buttons in the dark, until the automobile connection was made. She could be found wherever the patient advocates gathered, making sure everything was all right for them.

On the last night, three champions spoke: Sally Temple, Bob Klein, and Shinya Yamanaka.

Sally Temple is genuinely beloved for her work to help the scientists of New York. She is currently President of the Neural Stem Cell Institute, and former President of many more organizations — including the ISSCR — but is a scientist first and foremost, currently working to find cures for blindness.

I wondered how Bob would handle his big speech — especially the questions which might follow.

Would he talk about the amazing accomplishments of CIRM?

For example: Did you know that the California Institute for Regenerative Medicine has funded more than 800 projects at over 70 Institutions? That it funded research for more than 75 different diseases? That it raised multiple billions of dollars in additional revenues for the state, as well as 25,816 new jobs.[5] And that it saved the lives of more than 50 children, and is helping develop treatments for diabetes, HIV/AIDS, leukemia, blindness, paralysis, cancer and more?

---

[5] http://tinyurl.com/y22ozmhg-

Or would he burst out with a huge revelation?

But no, Bob did not talk about what everybody was thinking — there was no hint of a 2020 initiative, and what would be the greatest stem cell battle of all time.

Instead, Bob spoke about the single most important help a patient advocate can do — to reach out to the media. The diminishing coverage of science is a subject that has worried him for years. CNN deleted their entire science department. 90% of science reporters have recently lost their jobs. Bookstores and newspapers are few and getting less. But we, the patient advocates, still possess the ability to shape the equation, if we will choose to do so. We are the most trusted group, because our motives are clear. We want cures for our loved ones. That's it. People understand that as a pure and clean motivation. But if we sit idly by, we doom our chances. Bob spoke for ten minutes, not a wasted word — and took no questions.

The final speech was made by Shinya Yamanaka, winner of the Nobel Prize for his development of a "substitute" for embryonic stem cells: induced pluripotent stem cells (iPSC). I put "substitute" in quotes because I am not convinced anything can take the place of hESC. We need both, and always will.

He began with a request for scientists to come to work with him in Japan. It would be ideal if they spoke both Japanese and English, he said, but not mandatory.

And then he got to the meat of the talk: a summary of what he had been working on with an entire country behind him. Although he did not say so, it is a fact that Japan is immensely proud of Dr. Yamanaka, and backs his work all the way.

He described clinical trials going on right now, using his iPSC invention, including: Age-related Macular Degeneration, Parkinson's disease, and Spinal Cord Injury. (As he once told Roman, Dr. Yamanaka began his scientific career as a student of paralysis.)

There was more, of course, as he told us what could be done with iPSC: "drug screening, toxicity testing, development of preventive measures..."

But something else came across as well, a gentle but powerful message about the purpose of science.

Wherever we live, across the globe, our bodies are identical: the same diseases afflict us all.

This warm, kind man, officially chosen the greatest scientist in the world, spoke often and openly, sharing what he had developed, what worked and didn't.

If we follow that example, sharing knowledge with each other, we have a chance to complete the puzzle of cure.

66 The Most Terrible Disease

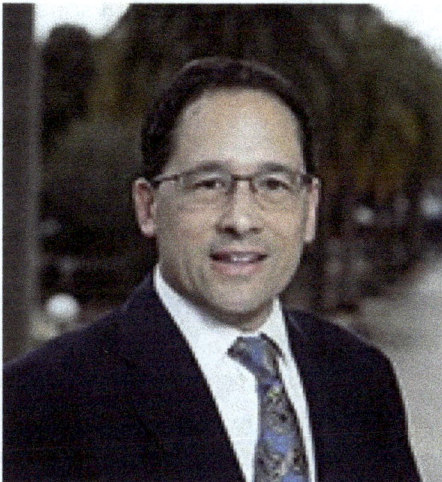

Stanford's Anthony Oro is challenging the most savage disease of all. (CIRM photo.)

What if you had a magic lantern with a genie inside, and he popped out and said <u>you could cure one disease</u> — which would you choose?

Like all good magic stories, the wish comes with conditions. You could not say, I wish an end to all chronic disease, nor for CIRM to have endless funding, (although a great wish!) — and above all, it could not be a condition that affects you or your family personally — what would it be?

Although there are hundreds of conditions deserving cure, still I know my choice.

Consider your *skin*. The largest organ in your body, skin can wrinkle, tan or sunburn, pick up an occasional pimple — nothing major — why even think about it?

Maybe because for most of us, skin does not split apart at the slightest touch.

But if you have Epidermolysis Bullosa (EB), a shortage of the natural skin glue (collagen) there are no guarantees of that.

Newborns with EB may come out into the world with their feet actually bleeding: torn by the pressures of birth.

For EB sufferers, eating a potato chip can rip the throat; rubbing an eye can send you to a dark room for days, or blind you. Scrubbing with a washcloth can pull the skin off. Slow-healing wounds may cover your body. Bandage changing may need to be done several times a day, families go broke paying for care — and EB patients die of squamous cell cancer.

John Hudson Dilgen is a young man whose photographs made the condition real to me. He was wearing what I thought were white sweat clothes, but they were not; they were bandages. His website gave me a hint of what EB means.[1]

Several days ago, CIRM stopped giving out any new grants. The money is essentially gone. Unless we can renew the funding, CIRM is done. (See next chapter for more on that.)

But grants begun will be paid out: promises will be fulfilled.

Dr. Anthony Oro of Stanford has a $5 million dollar grant to work on EB now. As long as that money lasts, the progress will continue.

There are two advantages to the grant.

First, for EB sufferers. The hoped-for outcome will be sheets of living tissue which cling on to the open wounds, and close them, heal them. That is the first goal, to end the suffering of EB patients.

There may also be another, more universal, benefit.

Here is how it might work:

First, the technique used to develop the sheets of living tissue is called induced Pluripotent Stem Cell research (iPSC), Yamanaka's Nobel prize contribution.

But not every stem cell line developed by iPSC is identical.

So what Dr. Oro is working on is a tool to predict which stem cell line will be best for a particular use; might not that technique be useful in choosing the best cell lines — from many other diseases?

Here is Dr. Oro:

"Induced pluripotent stem (iPS) cells are derived from skin that...can be coaxed into many different types of cells...and also back to skin cells.

---

[1] http://johnhudsondilgen.com/life.html

"One recognized...difficulty is that every individual iPS cell line behaves slightly differently, and protocols...refined to work well in one particular line do not work as well for another IPS cell line...this variability represents a major stumbling block for...a therapeutic approach.

"We propose to develop a tool to...*identify lines that will work well* in a given differentiation protocol...specifically...whether a given iPS cell line will be able to...give rise to skin cells.

"...(This may) make the tool usable...for predicting differentiation of iPS cells *into other medically important issues...*"[2]

Imagine yourself as a scientist with several hundred slightly different stem cell lines — which one is best? This tool might help you choose.

Techniques developed to fight EB might apply to other conditions — and bring relief to many.

---

[2] https://www.cirm.ca.gov/our-progress/awards/chromatin-context-tool-predicting-ips-lineage-predisposition-and-tissue

67 Body as Battlefield: Clinical Studies Funded by CIRM

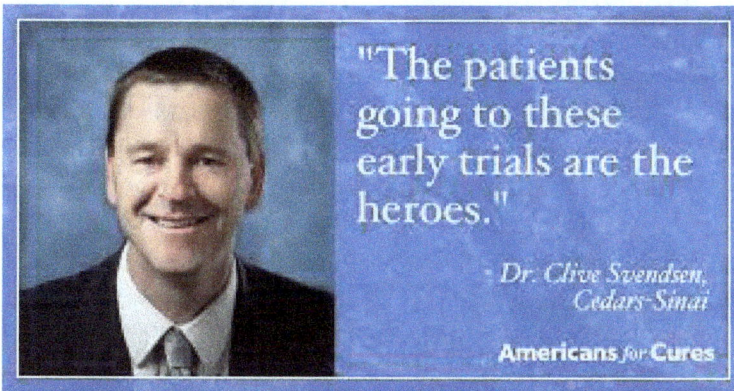

"The patients going to these early trials are the heroes."

Dr. Clive Svendsen, Cedars-Sinai

Americans for Cures

CIRM has 54 human trials underway as this is written. Clive Svendsen is leading one of these trials, an attempt to "slow down the progression of ALS" (Lou Gehrig's disease). (Americans for Cures Foundation photo.)

On the desk beside me is an 11" by 17" document — heavily censored. *There is nothing bad in it (just the opposite, it's wonderful!), but it is a lot of material, mostly in the giant words of science. I shortened it.

You are about to encounter actual human trials, funded by CIRM.

To reach human trials is the goal of the program. Once we know what therapies actually help people, then the private companies can step in and develop the products, make them available to the public. (Remember, by California law, CIRM itself cannot own or sell the treatments.)

Some thought we would be lucky to reach even <u>one</u> human trial, due to the extreme costs and difficulty of testing. But we did — **51 HUMAN TRIALS**.

AND — as of October 31, 2019, the latest number is — 60![1]

---

[1] https://www.cirm.ca.gov/clinical-trials

This is the body as battlefield, inside which diseases are being fought. Some procedures were successful, some not. Some of the people in the trials were literally dying when they entered. Some got better; some did not.

Every project began as somebody's dream; most of which faded in the morning light. These are the survivors. They endured years of testing: first as cells, then some with weird stuff like the larvae of flies, or zebra fish, mice, rats, pigs, monkeys, sheep — until, if all goes well — human trials.

For each trial listed below, I have included: the project number, a scientist, test site, disease (underlined), and maybe some interesting bit of information.

Materials were gathered by **Mitra Hooshman**, PhD, Chief Science Officer for Americans for Cures Foundation, and **Mary Bass**, former Executive Director, Americans for Cures. Huge chore! Results were last updated October 31, 2019. I have only cited information for the first 51 trials.

Note: errors are my fault, and will, if noted, be corrected in future editions.

So — here we go!

1. CLIN 2-10144, Mark Chao, Forty Seven, <u>Acute Myeloid Leukemia:</u> "Cancer cells have a "don't eat me" signal on their surface that makes them unrecognizable by immune cells. The (product antibody)... identifies the "don't eat me" signal and binds to them, marking them for destruction."

2. CLIN2-09339-A, Donald Kohn, UCLA, <u>ADA-SCID</u> (**"Bubble-Baby Syndrome"**): "42 babies cured (now more than 50! — DR) and therapy may accelerate (by) FDA approval."

3. CLIN2-08289, Mehrdad Abedi, UC Davis, <u>AIDS-Lymphoma</u>: "HIV stem cell gene therapy (may) replace a patient's (faulty) immune system."

4. CLIN2-09183, Tippi McKenzie, UCSF, <u>Alpha Thalassemia Major:</u> "...in utero transplantation (of cells) could result in a definitive cure, or allow postnatal transplant with decreased morbidity. The first procedure has already been successfully completed, and led to the birth of a child whose morbidity likelihood was over 90% otherwise."

5. CLIN2-09894, Ralph Kern, Brainstorm, <u>**ALS**</u>: "(In) Israel and the US, 98 patients (had) transplantations in phase 1–2 clinical trials...The... results of these two trials...show an average 77% increase in lung function..."

6. CLIN2-09284, Clive Svendsen, Cedars-Sinai, <u>**ALS**</u>: "...New astrocytes...have been modified to release (a growth factor)... to dying motor neurons, (which, it is hoped) will slow down the progression of ALS."

7. DR3-07438, Mark Humayun, USC, **<u>Age-related Macular Degeneration</u>** (AMD): "...embryonic stem cell-derived cells are delivered as a thin layer (on the back of) the eye... 4 of 5 subjects successfully received the implant...none showed progression of vision loss. 1 eye improved by 17 letters, 2 eyes improved fixation."

8. CLIN2-10830, Morton Cowan, UCSF, <u>**ART-SCID**</u>: "using gene-corrected cells from the patient should eliminate SCID and resume normal immunity for children affected by the disease."

9. CLIN2-10386, Paul Finnegan, Angiocrine, **<u>Blood Cancers</u>**: "With many "expansion" technologies, growing more stem cells causes them to lose their quality...This (might preserve more of the cells' potency? — DR).

10. CLIN2-11371, Edward Kavalerchik, Angiocrine, **<u>Blood Cancers</u>**: "AB 205 consists of Human Umbilical Vein Tissue...this therapy could improve quality of life and reduce the need for high dose chemotherapy/radiation."

11. CLIN2-10846, Crystal Mackall, Stanford, **<u>Blood Cancers</u>**: "Upon recognition, the T cells will become activated, divide, and then kill the cancer cells... T cells within the larger population will form memory stem cells that will... continue to survey... and kill residual cancer."

12. DR3-06965, Irv Weissman, Stanford, **<u>Blood Cancers</u>**: (see # 1): "62 patients with advanced tumors treated. Preliminary evidence of anti-tumor activity observed in patents with ovarian cancer.

13. CLIN2-08280, Anthony Gringeri, Immunocellular Therapeutics, **<u>Brain Cancer</u>**: "... T cells that specifically target...tumor cells... could provide a selective means of ablating cancer cells...otherwise resistant to chemo..."

14. CLIN2-09574, Colleen Delaney, Nohla Therapeutics, Chemotherapy-induced **<u>Neutropenia</u>** in AML patients: NLA101 is an off-the-shelf,

universal donor line... Upon transplantation into patients who have received high-dose chemotherapy, NLA101 is thought to reconstitute white blood cells effectively and reduce neutropenia.

15. CLIN2-09577, March Chao, Forty Seven, **Colon Cancer**: the antibody has "been used in clinical trials targeting cancer cells, including Hodgkin's and non-Hodgkin's lymphoma...it was safe and is under phase 2 testing."

16. CLIN2-08334, Deborah Ascheim, Capricor, **Duchenne's disease**: "to test the efficacy of...stem cell transplantation in Duchenne Muscular Dystrophy on improving heart muscle regeneration... 89% of the... treated patients...demonstrated sustained or improved muscle function..."

17. DR2A-05735, Rachel Smith, Capricor, **Heart Failure**: Treated 134 patients w/CAP 1002 or placebo. Interim analysis determined that trial unlikely to reach pre-defined efficacy endpoint..."

18. DR1-06893, Geoff Symonds, Calimmune Inc., **HIV/AIDS**: "Objective: To complete a phase 1–2 study for use of gene-modified (human stem cells) for HIV treatment...(Unfortunately): Calimmune was acquired by CSL Ltd. in 2017 and CSL has not continued with the study..."

19. SP3A-07536, John Zaia, City of Hope, **HIV/AIDS**: Objective: "To complete a phase 1 study for treatment of HIV using zinc finger nucleases...a molecular scissors — to snip out the target gene... First 3 participants have had the gene-modified blood stem cells manufactured and successfully treated...Trial continues with enrollment of more patients..."

20. DR2A-05415, Vicki Wheelock, UC Davis, **Huntington's Disease**: Objective: "To complete a phase 1 study for treatment of Huntington's disease, using genetically-modified... stem cells... Observational clinical study (NCT01937923) was (funded by CIRM and) completed. However, CIRM did not fund (the next step)..." (But — good news from Jan Nolta of UC Davis:) "Fortunately we submitted the same grant application to the NIH and got a perfect score, so it was funded.")

21. CLIN2-10344, Damien Bates, SanBio, Inc., **Ischemic stroke**: Objective: "to complete phase 2 trials using bone marrow-derived stem cells...Impact: 16 patients completed a phase 1–2 trial, and at 12 months post-transplant, ... had increased motor improvement and reduced...damage..."

22. CLIN2-09439, Samuel Strober, Stanford, **Kidney Disease**: "The Stanford protocol (is) efficacious in... kidney transplantation, allowing cessation of immune-suppressive drug therapy in 80% of study participants..."
23. CLIN2-08938, Jeffrey Lawson, Humacyte, **Kidney Failure**: "Objective: to complete a phase 3 trial using...smooth muscle cells...for patients undergoing kidney failure and not suitable to receive dialysis. The... vessel is not meant to treat kidney failure itself; rather it (fits) into the arm of the patient and enables dialysis...60 patients transplanted (40 in Poland, 20 in U.S.). No evidence of immune rejection..."
24. CLIN2-09688 (same as 23, except no longer recruiting) **Kidney Failure**.
25. CLIN2-11400, Everett Meyer, Stanford, **Kidney Failure**: "Study would allow for (no) immune-suppression... (during) kidney transplants."
26. CLIN2-10411, Scott Barry, Medeor Therapeutics, **Kidney Transplants**: "...eliminated need for all immunosuppressive drugs in 23 of 28 patients..."
27. DR3-06924, Thomas Kipps, UCSD, **Leukemia/Blood Cancers**: "To complete phase 1b–2a trial for cirmtuzumab when given alone or in combination with ibrutinib to patients with leukemia and/ or lymphoma...proposed to significantly increase proportion of patients with complete remission and long-term cancer control..."
28. CLIN2-10192, Thomas Kipps, same as # 27, Phase 2, no results as yet.
29. CLIN2-10784, Steven Dubinett, UCLA, **Lung Cancer**: "...to genetically modify a patient's own...cells with an adenovirus to...stimulate native T cells to destroy cancer cells. Combined with FDA-approved pembrolizumab...(may) result in the inability of the cancer cells to 'hide'"...
30. CLIN2-10248, Christine Brown, City of Hope, **Malignant Glioma**: "...utilizes patient's...T cells to...destroy malignant glioma..."
31. CLIN2-08239, Robert Dillman, Caladrius, **Melanoma**: "...using the patient's own tumor cells to create a personalized therapy, to... destroy the cancer... Phase 3 terminated...due to financial reasons. (NCTO3241875653)"
32. DR2A-05309, Antoni Ribas, UCLA, **Melanoma**: "Genetic modification of human stem cells with a gene, NYESO-1, (which recognizes and marks cancer cells for destruction)... first patient has been transplanted."

33. CLIN2-10395, Mathew Spear, Poseida Therapeutics, Inc., **Multiple Myeloma**: "Autologous T-cells are collected and... modified with a receptor that...binds PBCMA, a protein expressed in most mature B cells in multiple myeloma. 23 patients... all... showed improved markers of disease..."

34. CLIN2-10388, Kristen Johnson, Calibr, **Osteoarthritis** (OA): "KA34, a (bone-growth enhancing) molecule...promotes cartilage (growth)... leading to repair of damaged cartilage of OA patients..."

35. DR2A-05302, Nancy Lane, UC Davis, **Osteonecrosis**: "reactivation of endogenous MSCs to generate bone..."

36. CLIN2-09444, Michael Lewis, Cedars-Sinai, Patients with **Pulmonary Arterial Hypertension (PAH)**: "CAP 1002 cells reduced wall thickening of small blood vessels in the lung that are...obstructed in PAH, (thereby lightening) the load on the heart, so it does not need to work as hard..."

37. CLIN2-09698, Henry Klassen, jCyte, **Retinitis Pigmentosa (RP)**: "Retinal progenitor cells are delivered to the back of the eye to replace lost photoreceptors... (of) 28 subjects, the average improvement in the treated eye was 10 letters on the BCVA test... for the visually impaired...

38. DR2A-05739, Henry Klassen, UCI, **Retinitis Pigmentosa:** a phase 2 trial using human fetal-derived retinal progenitor cells. "This study completed its first phase and went on to a phase 2 trial that is also now complete."

39. CLIN2-10847, Joseph Rosenthal, COH, **Sickle Cell Disease**: "(achieving) the right mix of donor-to-host blood cells may reverse sickle cell disease. Immune tolerance will prevent rejection of the donor blood stem cell graft and allow patients to be free of sickle cell disease for a long time."

40. DR3-06945, Donald Kohn, UCLA, **Sickle Cell Disease**: "... modified gene/stem cell transplant in Sickle Cell patients...Similar to the treatment of ADA-SCID ("bubble baby syndrome"), this therapy could potentially cure sickle cell disease...

41. DR3-07067, Dennis Slamon, UCLA, **Solid Tumors**: "... treatment of cancer...block(s)...a protein important for survival of the cancer stem cell...may stop or even shrink tumor growth... could provide a selective mechanism for destruction of cancer cells..."

42. CT1-05168, Jane Lebowski (Geron) **Spinal Cord Injury**: "a phase 1 trial using hESC-derived oligo-progenitors for treatment of spinal cord injury. (Note: this is the famous embryonic stem cell research first funded by a Roman Reed grant, and performed by Hans Keirstead, rats that walked again after being paralyzed, as seen on TV's 60 MINUTES.)… Patients were safely transplanted and IP rights of study…were turned over to Asterias Therapeutics. (For financial reasons, Geron left the stem cell business — DR).

43. SP3A-07552, Ed Wirth, Asterias Therapeutics, **Spinal Cord Injury**: "to carry out a phase 1 trial (more cells than in 42) using hESC-derived oligo-progenitors for treatment of spinal cord injury. 25 of 25 patients were safely transplanted… over 80% of patients gained 1–2 levels of motor recovery."

44. CLIN2-11031, Ed Connor, Sangamo, **Beta-Thalassemia**: "a gene-edited therapy candidate for patients with transfusion-dependent beta-thalassemia…ST-400 may potentially reduce or eliminate need for chronic blood transfusions and improve patients' quality of life…"

45. CLIN2-09730, Douglas Losordo, Caladrius Biosciences, **Type-1 Diabetes**: "to complete phase 2 (safety) trials using… expanded T-regs (cells). Blood is drawn, T-regs are isolated and purified… transplanted back…safety…has been shown in 22 patients and… even evidence of increased remission"

46. CLIN2-09672, Howard Foyt, ViaCyte, **Type-1-Diabetes**: testing of VC-02, Pec-direct, Encapsulation of hESC-derived pancreatic cells. "…study currently recruiting for phase 1–2. 40 patients (will be) in Cohort 2 of trial. First patients implanted with potentially efficacious dose of PEC-direct… Primary efficacy measurement… clinically relevant production of insulin".

47. AP1-08039 and SP1-06513, Howard Foyt, ViaCyte, **Type-1-Diabetes**: "…transplanted cells, delivered in a device, are thought to be protected from immune attack, while differentiating into pancreatic cells to regulate blood sugar levels. Consistent, robust engraftment limited in this study but when it did occur, viable mature *insulin-expressing endocrine islet cells formed, some persisting for up to two years after implantation.*" (emphasis added).

48. CLIN2-10392, Michael Pulsipher, CHLA, **Viral Infections**: "(a) virus-specific T-cell therapy targeting cytomegalovirus, Epstein–Barr virus,

and adenovirus... viruses account for up to 40% of deaths in patients with immunodeficiency... Restoration of T-cell immunity...could provide lasting control of targeted viral infections."

49. CLIN2-08231, Donald Kohn, UCLA, <u>x-linked **Chronic Granulomatous Disease**</u>: to complete phase 1/2 trials. "... correct(ions) for the missing gene in this disease will be transplanted into the patients and (hopefully — DR) will lead to production of healthy white blood cells that have the capacity to kill infectious microorganisms...Trial is currently ongoing and could cure patients from this disease." (Note: it has; see Brenden Whitaker.)

50. CLIN2-09504, Stephen Gottschalk, St. Jude, **X-SCID**: "X-SCID (is) a fatal pediatric disorder... For treatment of X-SCID, a "similar approach to Don Kohn's (is recommended)...to deliver a normal copy of the...gene that is defective. Correction of the immune cells will allow the patients to fight infections that are life-threatening — without curative therapy." The patients will be spared life-long (expensive treatments of other kinds).

51. DR2A-05365, Judith Shizuru, Stanford, **X-SCID**: "Objective: to assess the safety and preliminary efficacy of a...chemotherapy-free conditioning for...transplants in patients with (SCID) Severe Combined Immune Deficiency... This could be applied to any other transplantation setting...(to provide) a safe and non-invasive method to enable transplant acceptance. NOTE: ... LACK OF FUNDING may block treatment of 6TH patient...

Flash: CIRM came through with a $3.7 million grant so Dr. Shizuru could complete her trial![2]

---

[2] https://www.cirm.ca.gov/about-cirm/newsroom/press-releases/02212019/stem-cell-agency-invests-chemotherapy-free-approach-rare

68 More Victims Than Five Kinds of Cancer?

Brian Cummings of UCI applied for a grant to fight Traumatic Brain Injury. The only catch? No money left in that category .... (Photo by UCI.edu.)

After funding stem cell research for 16 years, CIRM's money was almost gone. There were "funding requests for $88 million...but only about 1/3 of that amount left to give out to scientists — and then there would be no more."[1]

On July 24, 2019, there would be a meeting of the board, to decide where the last money would go.

America's greatest source of stem cell research funding would soon be shut down.

Important research questions would go unanswered, because there was no money.

Example: a grant request to fight traumatic brain injury, generally considered impossible to cure.

---

[1] https://www.the-scientist.com/news-opinion/stem-cell-funding-agency-cirm-is-nearly-out-of-funds-66108

Brian Cummings of UC Irvine was the Primary Investigator. He asked my advice, and naturally I gave it to him — as comedian Alan King once said, "Free advice costs nothing, and it's worth the price!"

My free advice was to come to Oakland, be there at the meeting, and fight for his grant.

Too many scientists think they can just mail in their grant requests and that it is the end of it. It may well be the end — of their hopes for funding!

I read Dr. Cumming's project several times and it seemed solid to me. Also, I knew three of the scientists who were working with him — Aileen Anderson, Hans Keirstead, and Gabriel Nistor — three of the best researchers in the world.

And so I also gave him my recommendation, which follows below.

Support for Traumatic Brain Injury Project Tran1
Public Comment by Don C. Reed:

Will there be a Prop 71 part 2? That decision is not mine to make. But I absolutely know what I do want, and that is a major renewal of funding.

The best way to move toward that goal, I believe, is to take on a truly huge problem — and they just don't get any bigger than Traumatic Brain Injury, TBI.

Every year, more than 200,000 Californians receive a traumatic brain injury, at a financial cost of roughly $9.6 billion, an amount exceeding California's entire ten year investment in CIRM, roughly $6 billion, including interest.

Across our country, 1.7 million citizens suffer a TBI — at the staggering expense of $76.5 billion dollars.

More people have a traumatic brain injury than are affected by cancers of the brain, breast, colon, lung and prostate — put together.

What is it like to have a traumatic brain injury? Often compared to Alzheimer's disease, TBI destroys memory, and brings emotional confusion to the sufferer. While TBI affects similar numbers of Californians with Alzheimer's, TBI is much less known, and might be called a silent epidemic.

The Primary Investigator for this project, Brian Cummings, told me of a family summer camp for children to which he brought his daughter. While there, he met a woman who had been a soldier in the Iraq war, where she twice received TBIs from one of those ghastly home-made bombs, the Improvised Explosive Device, an IED. What brought the meaning of Traumatic Brain Injury home to Dr. Cummings was that this woman soldier — could not remember which were her children.

Traumatic Brain Injury, at present, is incurable.

As you know, for more than 25 years I have been supporting research for the related condition of spinal cord injury. And four of our greatest research champions are: Aileen Anderson, Hans Keirstead, Gabriel Nistor and Brian Cummings. All four of these outstanding SCI scientists — this time led by Brian Cummings as the TBI expert — will be involved in this project.

Their goal? "Transplantation of human neural stem cells could lead to improvements in learning, memory and emotion (to) significantly change a patient's quality of life, (with) considerable economic impact to California."

In addition, four young scientists from CIRM's Bridges program will bring the energy and passion of youth to this endeavor.

This is CIRM at its very best. I urge your support.

Sincerely,
Don C. Reed

And then there was nothing more I could do, except wait until July 24th, (day after tomorrow!) and see what CIRM's decision would be.

69   Vertigo, Chickens, and Maybe Great News

Could we use something connected to seasickness to restore hearing to the deaf? Alan Cheng is the man to ask. (Photo by Stanford.edu.)

In Alfred Hitchcock's classic thriller, "VERTIGO," James Stewart suffered attacks of the title-named condition as a central part of the plot. He had to climb up high buildings again and again, putting him into intense misery from his loss of balance, and the director wanted the viewer to share his discomfort.

But how to make visual the feelings of a vertigo sufferer? Hitchcock invented a technique called the "dolly zoom," where the camera holder moves back and away from the object he/she is filming. The result is a sudden shock of confusion, as if the world was yanked out from under.

For viewers of the movie, this was a brief intense movie scene.

But what if that stomach-clutching dizziness was something you had to live with on a permanent basis?

Dizziness, like deafness, is caused by problems in the inner ear.

Deep inside the ear are two tiny pools of water: one having to do with balance, the other with sound. Sticking up from these pools are hair cells (stereocilia) which vibrate, sending information to the brain.

Haircells in the balance pool can sometimes regrow, those for hearing cannot.

What might that be like, to be permanently deaf or dizzy?

Once I visited a school for the deaf. And there was an indoor basketball game going on. I watched it and kept feeling something was missing. Then it hit me.

Aside from the occasional grunt of exertion, and the squeak of rubber sneakers on the floor, it was silent. No one talked or yelled because no one could hear...

The loss of hearing is so big a concept I cannot really grasp it.

But dizziness? That I can understand, at least a little.

When out on a boat and my face goes cold and pale, and I am leaning over the side, hoping for some quick relief such as extinction, some non-seasick person will inevitably advise, "Just don't think about it." Don't think about it? To overlook a massive headache, the world spinning around, plus your entire stomach wants to depart your body?

Even so, my balance problems are minor, and temporary. If I am going to be on the ocean, I put one of those seasick patches behind my ear, and I am pretty much okay. (Gloria never gets seasick at all, which I find grossly unfair.)

Some, however, may become so debilitated by dizziness that they are bedridden...

How many people suffer chronic balance disorders like vertigo? Estimates vary wildly. But it would appear to be about eight million, according to the National Institute on Deafness and Other Communication Disorders (NIDCD).[1]

How many people are either deaf or hard of hearing? An astonishing 48 million...[2]

---

[1] https://qz.com/909898/getting-diagnosed-with-a-chronic-disease-with-no-treatment-or-cure-is-like-entering-a-kafka-novel/

[2] Lin FR, Niparko JK, Ferrucci L (2011) Hearing Loss Prevalence in the United States. *Arch Intern Med* 171(20):1851–1852. https://www.ncbi.nlm.nih.gov/pmc/articles/PMC3564588/

CIRM-supported Alan Cheng is working on both conditions: balance problems like vertigo as well as hearing loss.

If you read my last book, "CALIFORNIA CURES", you will remember Stanford's Dr. Alan Cheng. An intensely creative scientist, Cheng's initial approach appeared to focus on the difference between hearing hair cells and balance hair cells.

Human hearing cells do not regenerate, but balance hair cells do. Perhaps by studying the balance cells, we could find a way to regenerate those for hearing.

If we could "teach" our hair cells to regenerate, we might actually reverse deafness, and restore the gift of hearing.

This would require working with something called the Wnt pathway: "a group of signal transduction pathways which begin with proteins that pass signals into a cell."[3]

Dr. Cheng wants to "modulate the Wnt pathway to restore inner ear function."

If a particular protein (Atoh1) was put into the balance portion of the inner ear (the utricle) which regenerates hair cells a little, a more reliable return would be possible.[4]

Working with Zahra N. Sayyid, as well as Tian Wang, Leon Chen, and Sherri M. Jones, Dr. Cheng found that:

"Atoh1 overexpression enhances regeneration and leads to a sustained recovery of vestibular (balance) function."[5]

But, Cheng adds in a personal communication: "Using Atoh1 is not enough to restore hearing."

And as I wait for Wednesday, July 24 CIRM's meeting, I think of all the work that will not get done, when the funding is no more.

---

[3] https://en.wikipedia.org/wiki/Wnt_signaling_pathway
[4] Sayyid ZN, Wang T, Chen L, *et al.* (2019) Atoh1 Directs Regeneration and Functional Recovery of the Mature Mouse Vestibular System. *Cell Rep* 9;28(2):312-324.e4. https://www.ncbi.nlm.nih.gov/pubmed/31291569
[5] https://www.cell.com/cell-reports/pdf/S2211-1247(19)30794-6.pdf

70 A World without CIRM?

Can you imagine a world without CIRM — without the California stem cell research program? That does not make sense to me.

I'll tell you what does make sense: a song the workers sang on their way to the factories in World War Two — and it went like this: "We did it before and we can do it again, yes, we can do it again — we did it before — and we'll do it again!"

I remember when we were first working on the state ballot initiative that created CIRM, in April 2003, and I had just met Bob Klein. He introduced me to a room full of patient advocates just like you who are reading this now.

There was a pile of cardboard boxes, and my first job was to take a razor blade and cut them up, to make backing for the signature sheets. I went home with a blood blister on my thumb and contentment in my heart, because I had done a little work toward our goal. That is the secret, of course, do something every day.

The results were fantastic. And it started from the top. When friends approached Bob Klein, to ask him if he would he raise $1 billion for stem cell research, he said "No!" Too much? Not enough, he said, it has to be at least $3 billion. He is always like that: a visionary who learns from experts what is needed, and then finds a way to empower them.

Even then, the opposition tried to shut us down, using multiple lawsuits — even one on behalf of a non-existent embryo. Fortunately, the initiative (Prop 71) was strong and well written. We prevailed in the courts and CIRM was able to start using the voter-approved bond funds for research grants. It took almost three years, but at last we were fully functional.

Today, some of us almost take CIRM for granted, because it does so much with no fuss, providing grants for scientists to tackle diseases long considered incurable.

Think of whatever disease or disability interests you. Chances are, CIRM-funded scientists are fighting it right now.

Example: Over 50 little boys and girls were born with no immune system — before CIRM, they might have lived inside plastic bubbles, or died — but not now! Today, they run around outside, exhausting their parents — as children are meant to do.

Example: If there was no CIRM — blindness would be far more likely to remain a permanent condition — today, the possibility of stem cell therapies for both the blindness affecting the old (Age-related Macular Degeneration) and the young (Retinitis Pigmentosa) are being tested in CIRM-funded clinical trials.

Example: Bladder cancer. Right now, the only way to handle this deadly threat is to remove the cancerous bladder, and replace it with a bag hanging out of your side. What if, instead, we could regrow and replace the inner lining, so the cancer was controlled and health restored? Thanks to a CIRM-funded grant, that possibility can now be investigated.

But what if the CIRM funding stopped? Imagine riding the elevator to the 16th floor of that building in Oakland where CIRM is headquartered, and we pushed the button on the office door, and nobody answered — ever again? We cannot allow that.

What will we do? The same thing we always do, as advocates. We will have our personal stories so polished and ready that we can tell them at the drop of a hat — our stories are the most important tool we have.

We tell about our own disease or disability, or a loved one's, and how CIRM-funded scientists are fighting it.

Like that my son Roman Reed was paralyzed in a college football accident — but today the California stem cell program has funded a therapy that has been proven safe and has improved function for many of the 25 newly-paralyzed people participating in the clinical trial.

We must succeed once more in California with our 2020 initiative to renew funding for CIRM. Unfortunately, in Washington, there are powerful folks who oppose some of the research we use. They want scientists to work with one hand tied behind their backs, using only the ideologically-approved techniques they want to see funded.

With CIRM, nobody asks about the political desirability of a particular experiment. The "ask" is always: what is the best way to save lives and ease suffering, within the ethics and laws of state and nation?

I recently heard leaders of two political parties, Joe Biden and Donald Trump, both saying they intended to defeat cancer — well, they should come to California and walk with Bob Klein through the halls of CIRM, and meet some of the scientists who dedicated their lives to the battle against cancer.

We must provide our scientists the funding they need: to conquer the chronic diseases which sweep our planet like a plague. Without funding, the best scientists can do nothing. Funding to them is like the weapons of war Winston Churchill asked for, saying: "Give us the tools and we will finish the job."

Before us lies the greatest stem cell battle of all time. You are its leaders. You do not need to get psyched up; but you do need to know you are not alone. Be organized with us. Stay active, stay in touch, do something every day — and we will get the job done.

Thank you in advance, for the hard work that lies ahead.

We must never let there be a world without CIRM.

71 What We Must Do

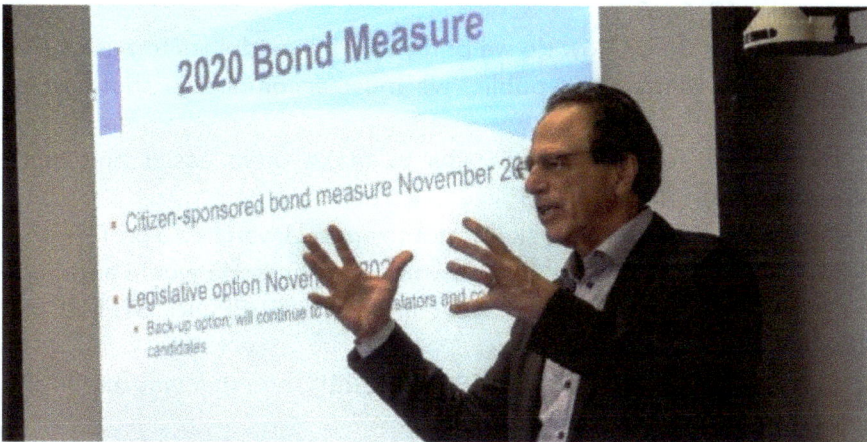

Bob Klein envisions a better future: can the patient advocate community help him make it become real? (Photo by David Jensen, Stem Cell Report.)

This book contains a small insider's joke, its number of chapters: 71. This is a reference to our citizens' initiative, Proposition 71, which became CIRM: the California Institute for Regenerative Medicine.

How good is CIRM? One clue is I have written not one but three books about it.

Great changes have begun. Steps are being taken toward alleviating blindness, cancer, kidney failure, paralysis and more — all being tested on people right now.

So the question is simple. Do we give up? Take our therapies and go home? Stick our research in some intellectual attic and forget about it?

CIRM must be protected, supported, encouraged, made safe. There is only one way to do this, and that is to renew CIRM's funding.

That means us, you and I. Wherever you live, whether you are old or young, struggling with a disease or healthy as a horse — the California Institute for Regenerative Medicine is for you, and those you love.

We fight to ease the suffering of our loved ones, to honor the memory of those who passed on, to protect our children, grandchildren, and generations yet unborn. Our enemy is chronic disease, and *that battle must go on.*

As Lincoln said:

"It is for us the living, rather, to be dedicated here to the unfinished work which they who fought here have thus far so nobly advanced." — November 19, 1863.

How do we do it? How do we bring $5.5 billion dollars to this cause?

When the battle goes public, we must accomplish a short list of tasks.

Here are a few, with some tips to make them easier and more effective.

Signature-gathering. The key is not to argue. I watched a professional do this, and he had the process down to *six words:* He would approach a person, offer his signature sheet and say: "SUPPORT STEM CELL RESEARCH?"

If "Yes," his next words were "SIGN HERE!", and that was that. If the answer was "No," he just moved on.

Telling Your Story: Tell it over and over again, no matter how bored you may feel repeating it — sometimes I just hate the sound of my own voice — but give it energy, make it new to the listener.

Boil it down to a couple sentences, like in my case:

"My son Roman Reed was paralyzed in a college football game. Stem cell research offers a chance for him to get better, so we support CIRM, the California Institute for Regenerative Medicine."

Question: has your story been in the newspaper or on TV yet? If not, put it there; 70% of all the news in the media is planted, put there by someone like you.

How is that done? Write your story in one page. Take time to do a good job: describe your condition of interest, and clearly communicate that you think CIRM should be renewed. Read it to yourself, over and over. When you feel comfortable with it, call your local newspaper, or TV station, tell them you have a *local angle story to suggest,* and ask who should you talk to? Talk to whomever they refer you to, and wrap up your story by asking if you can send them information. They will say yes. You email them a 1–2 page version of your story, follow up in three days with a phone call. They will assign you a reporter.

When they run the article in the paper, make a bunch of copies. Now you have a newsclip or a URL — on paper or TV watchable — and you send that that out as background, whenever you write a letter of support.

We always need **Letters of Support**. Begin with one sentence:

"I (state your name) support the California Institute for Regenerative Medicine, because (I or your loved one) has (condition) and then — tell your story! Be clear we are fighting for real people — and there are a lot of us! Roughly one American in five has a disability — nearly one in two has a chronic disease. If we vote as a bloc, nothing can stop us!

Involve Your Social Groups: we need the endorsement/support of every social group, big or small (we had 80 last time): every medical organization, every political group, every scientist or patient group. Want to keep up with what's happening? Here are some suggestions:

1. Go to www.americansforcures.org and sign up; it costs nothing; all that will happen is that you now have a good source of information. Americans for Cures sprang from the campaign for Proposition 71. When the fight begins, this is where you will find out what is happening. Or: contact Field Coordinator Melissa King, at: mking@americansforcures.org

2. **Listen strongly to new friends.** Expect a new advocate to need an hour to tell you his/her story. These are the warriors, and they deserve to be listened to.

3. Go to my website, www.stemcellbattles.com and sign up for the free newsletters. Also, consider buying my earlier books, STEM CELL BATTLES, or CALIFORNIA CURES, at a discount at the publisher's website: https://www.worldscientific.com/worldscibooks/10.1142/9255, also available at Amazon or Kobo. Or ask for them at your local library.

4. Visit David Jensen's massive blog, the California Stem Cell Report. He is an honest reporter, and gives both sides. I read everything he writes, and he writes a lot. Valuable resource. http://californiastemcellreport.blogspot.com/

5. CIRM itself. Go to www.cirm.ca.gov, and lose yourself in its endless pages. Sign up for their weblog, written by numerous authors but particularly by Kevin McCormack. https://blog.cirm.ca.gov/ If you

want to see what CIRM does, come to some of the many public meetings; you are welcome.

But you think you are not qualified to do such work? Neither is anyone. We just do it. The decision makers on the ICOC board (the Governing Board) are wonderful people, champions all — but less than half of them are scientists. The other half are patient advocates, like you.

Join us! As my son Roman Reed always says, "Take a stand with us, in favor of medical research. Take a stand — so one day, everybody can."

If you want to reach me personally, drop an email to: diverdonreed@ gmail.com

Afterword

Why Advocates Fight

According to a recent RAND survey,[1] as many as 150 million Americans have one or more chronic illnesses; 100 million Americans have two or

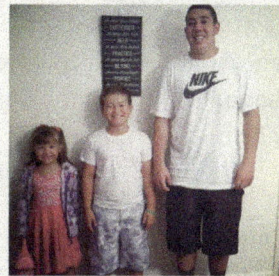

Katherine competing in muy thai (top left); Jackson, Desiree and Josh in Nevada (top right); Jason and Roman Reed politicking (bottom left); Roman's big three: Katherine, Jason, and "Little Man" Roman (bottom right).

---

[1] https://www.rand.org/blog/rand-review/2017/07/chronic-conditions-in-america-price-and-prevalence.html

more chronic diseases or disabilities; and almost 30 million Americans have five or more long-lasting or permanent ills. These are not empty statistics; but our families, your loved ones and mine, are all at risk; they are why we fight. Here are some of my personal inspirations.

First, good news about grants in question at the end of this book — Mark Tuszynski for Spinal Cord Injury ($6,235,897), and Brian Cummings ($4,833,271) for Traumatic Brain Injury — both received funding grants to move ahead.

Thank you for putting up with me and my family in the pages of these three stem cell books. Here is a quick roundup on how they are doing.

Gloria is still not recovered from the shoulder replacement surgery, but she is better. The pain is continual, but she is more functional every day. She succeeded in brushing her hair with one of those burning hot things — curling iron? — so she can make herself beautiful for herself again, though she already is, in my eyes.

Roman is strong, though at 46, he has many of the problems connected with spinal cord injury paralysis. But Roman is still Roman, and persists. There are a lot more surprises — political and otherwise — yet to come from Roman Reed.

Our daughter Desiree (remember she is Athletic Director for the University of Nevada at Las Vegas) is in town this week to run a 26-mile marathon.

Our grandchildren continue to give us joy: Jackson, Desiree's son, is a businessman and writer, getting paid for sports articles he writes from a young person's perspective. Roman's son Roman Jr., misnamed "Little Man" as he is 6'3" and built like a gorilla, continues his studies at the University of California at Berkeley. Jason still loves baseball, but lately has branched out into video games, at which he excels; and Katherine the great, who sends this message: "Don't pollute, don't overfish, save the ocean, save the Earth!"

From Bob Klein, a personal message:

"Ours to win, ours to lose; for the patients, the future is ours, if we act — united."

And from me, a hug: thank you for being a part of this great adventure.

Best always,

Don C. Reed

July 28, 2019

Name Index

Subject Index